Into the
NUCLEAR AGE

1925 TO 1945

Published by The Reader's Digest Association, Inc.

London • New York • Sydney • Montreal

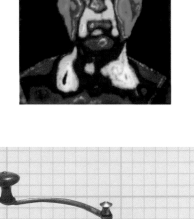

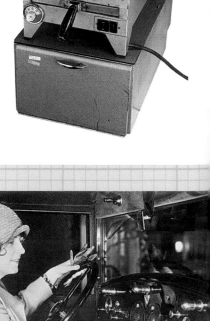

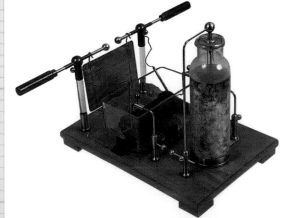

Contents

Introduction

The years between the wars were a time of major upheavals, many of them positive in their effects on quality of life. In a number of countries women won the right to vote; Ireland gained its independence, with the six counties of Ulster opting to remain with the United Kingdom; the leisure industries got into gear, with sport, radio and cinema all expanding fast. Yet these developments were overshadowed by menace. The interwar period was marked by the rising threat of movements that had their origins in the settlement that ended the First

World War and the humiliation inflicted on Germany. Ideologies sprang up that proved strong enough to displace older beliefs in God or in progress. Fascism, Nazism, Stalinism – with their charismatic leaders Mussolini, Hitler and Stalin – became the new opiates of the people. As the colonial empires consolidated, and Japan made its presence felt in the Far East, the world became increasingly divided into hostile blocs.

Though expressed ideologically, the origins of the Second World War lay in economics. The Wall Street Crash of 1929 set in train the Great Depression, putting millions out of work first in America and then around the globe, plunging the world into an unprecedented economic slump. Increasingly, the prospect of future conflict became the driver of technological innovation and industrial production. When war did break out, in 1939, it was on a scale never before known: a greater number of soldiers were involved in lands more far-flung than ever before, equipped with the most lethal weaponry so far conceived. The images still haunt us today – starving prisoners in concentration camps, entire cities razed by bombs into piles of debris. Most alarmingly of all, the Second World War saw the start of the nuclear age. Mankind had created its ultimate weapon and the means of its own potential destruction.   *The editors*

Manhattan project
In the early decades of the 20th century, New York – more specifically, the island of Manhattan – turned itself into the most modern, cosmopolitan city on the planet. The magnificent art deco Chrysler Building, seen here glinting in the sun (far left), was briefly the tallest building in the world, before being upstaged in 1931 by the Empire State Building, which held the title for four decades.

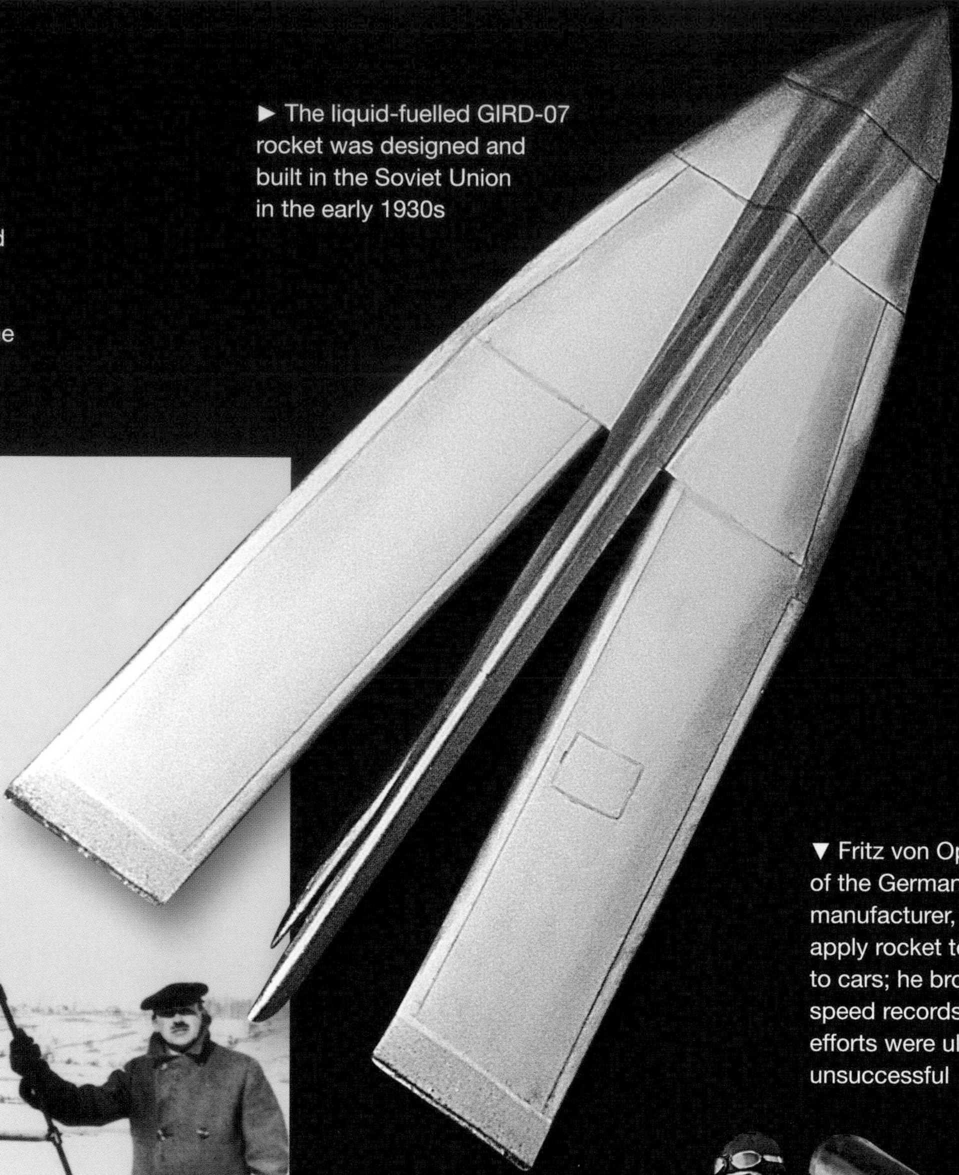

▼ The first successful firing
of a liquid-propellant rocket,
by Robert Hutchings Goddard
in 1926, was a major step on
the way to space exploration
but also opened the way to the
development of missiles that
would revolutionise warfare

▶ The liquid-fuelled GIRD-07
rocket was designed and
built in the Soviet Union
in the early 1930s

▼ Fritz von Opel, son
of the German car
manufacturer, tried to
apply rocket technology
to cars; he broke land
speed records, but his
efforts were ultimately
unsuccessful

The struggle for mastery of the skies had got under way in earlier decades, but the interwar years saw the race picking up pace. The invention first of the liquid-propellant rocket and then the helicopter offered new avenues for exploration. Charles

▲ Erwin Schrödinger, one of the founders of quantum physics, came up with his famous thought experiment involving a cat in a box in 1935; it was intended to illustrate the absurdity of some notions of quantum science when applied to daily life

◄▲ In 1925 John Logie Baird, a Scottish engineer of modest means, became the first person to transmit images over distance, but it was Philo Farnsworth (above and left), an American inventor, who was the first to develop an all-electronic television system without moving mechanical parts

Lindbergh made the first solo transatlantic flight, but non-stop commercial services across the North Atlantic were still some years away. Communications took the path trailed by radio before 1914, with television and talking pictures joining the advent of

▼ American astrophysicist Edwin Hubble formulated the law bearing his name that specifies that the distance of galaxies from Earth is proportional to the speed with which they are moving away from it, thereby demonstrating the continued expansion of the universe

► The introduction of talking pictures in 1927 opened a new era for cinema and a fresh generation of stars; from now on, audiences were able to hear as well as see the famous MGM lion

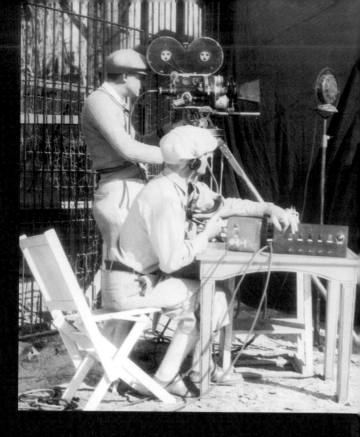

▼ On 21 May, 1927, US aviator Charles Lindbergh landed the *Spirit of St Louis* at Le Bourget airport in France, having made the first solo non-stop flight across the Atlantic in 33 hours and 30 minutes

mass tourism to contribute to the gathering leisure revolution. Cars continued to improve through innovations such as front-wheel drive. The launch of the Volkswagen Beetle, the first European 'people's car', was halted by the war, but the car would eventually

▶ Although quartz horological technology was initially rejected by all the major clock-making concerns, it eventually made highly accurate time-keeping affordable for all

▲ ▶ The antibiotic properties of penicillin, a mould that prevents the growth of certain bacteria, was first reported by Scottish bacteriologist Alexander Fleming in 1929; large-scale production of penicillin as a medicine got under way in the Second World War

outstrip the Model T Ford in sales. In physics, the revolutionary ideas first floated by Albert Einstein and Marie and Pierre Curie were now confirmed by Erwin Schrödinger's quantum theory and artificial radioactivity uncovered by Irène and Frédéric Joliot-Curie.

▲ ▶ The electron microscope, invented by the Germans Knoll and Ruska in 1931, revolutionised scientific observation, opening up new possibilities for exploring the world of the very small

▲ First spotted by the astronomer Clyde Tombaugh in 1930, Pluto was long considered the ninth planet of the Solar System, but was reclassified as a dwarf planet in 2006

Meanwhile, the discovery of Pluto by Clyde Tombaugh and Edwin Hubble's revelation of the expanding universe pushed back the boundaries of space. Yet the greatest number of innovations in the era were spurred by the looming threat of renewed war. Research

◄ In the 1930s, at a hospital in Copenhagen, a Professor Haxthausen adapted infrared photography for medical purposes; thermal imaging has since been put to use in many different fields

◄ The Hammond organ, patented in 1934, is named after its inventor Laurens Hammond

▲ The Central Asian Mission, which crossed the Asian landmass between 1930 and 1932, was masterminded by André Citroën as a publicity stunt to demonstrate the reliability of Citroën vehicles

on rockets led to the development of the first cruise missiles in the form of V-1s and V-2s, used by Germany against Britain in the latter stages of the Second World War, and of rocket-launchers like the Russian Katyushas. The first jet aircraft flew in Germany in

◄ Patented in 1930, glass fibre has many varied applications today, from an insulation material to uses in the optical and telecommunication fields to making car bodies and boat hulls

► A manual coffee grinder which won a prize at the Concours Lépine, a competition for inventors first staged in 1901 and held annually in Paris ever since

► In 1934 astrophysicists Walter Baade and Fritz Zwicky coined the term 'supernova' and put forward the theory that the intense brightness of the phenomenon is caused by an explosion that in fact represents the death of the star itself

1939, while parallel development of jet engines in Britain was spearheaded by Frank Whittle, paving the way for the post-war development of modern jet airliners. Some military inventions turned out to have applications in daily life, and vice versa: radar,

CINÉ-KODAK
KODACHROME
SAFETY
22° COLOR 22°
FILM

LENGTH 50 FEET WIDTH 16MM.
(15.24 meters)

MADE IN GREAT BRITAIN BY
KODAK LIMITED LONDON
TRADE MARKS PROTECTED THROUGHOUT THE WORLD

◄▼ Photography became a popular hobby in the 1930s, largely thanks to such inventions as Kodachrome film and the Leica compact camera

► Radar, which detects objects by bouncing radio waves off them, was developed in the 1930s; on the eve of the Second World War a line of radar stations, codenamed Chain Home, was set up along Britain's south and east coasts and would prove invaluable in the coming Battle of Britain

for instance, heralded the coming of microwave ovens, while nylon was used to make not just stockings but also parachutes; rubber found its way into inflatable boats as well as car tyres. Many scientists contributed to the war effort – none more so than

◄ The first artificial heart was made in 1937, but it was 1982 before Robert Jarvik (left) succeeded in implanting one in a human patient

▲ Cheap, robust and with a highly distinctive body shape, the Volkswagen Beetle, Germany's 'people's car', quickly conquered the world and remained a popular favourite for decades

▲ Helicopters became a viable proposition in the 1930s, but played only a limited role in World War II, becoming essential military equipment with the Korean War

Alan Turing who prefigured computers by devising a decryption machine, housed at Bletchley Park, capable of cracking German codes. Discoveries in the field of medicine helped doctors to control epidemics and provide much-improved care for the

◄ Although the 'leisure society' was still far in the future, the granting of paid holidays – in Britain, in 1938 – allowed people even on modest salaries to discover the joys of a vacation

► The Germans used their Enigma machine to encipher secret messages in the Second World War, but thanks to the work of Alan Turing in developing a decrypting machine, the Allies were able to break the German code, thereby gaining a decisive advantage

▲ Patented in 1938, nylon immediately proved adaptable to many uses, which in the Second World War included making parachutes as well as nylon stockings

wounded – this last one of the key positives to emerge from the war. Penicillin and cortisone entered the therapeutic arsenal, and the typhus vaccine did much to lessen the impact of one of the world's most lethal diseases, which raged anew in the wake of

► In 1932, with the help of this simple piece of equipment, James Chadwick demonstrated the presence of neutrons at the heart of the atomic nucleus, a decisive step in unravelling the atom

▲ On 5 August, 1945, the *Enola Gay*, an American B-29 bomber, dropped the world's first atomic bomb on the Japanese city of Hiroshima

► Developed in Britain and Germany in the 1930s, jet propulsion enabled aircraft to fly faster and higher than ever before

▲ Winston Churchill with a walkie-talkie – a portable, short-range radio transmitter-receiver that became a standard feature of World War II battlegrounds

the conflict. If any one city in the world can be picked out as symbolising the years between 1925 and 1945, that city was New York. It was there in the Roaring Twenties that millionaires competed to raise the highest skyscrapers. It was the scene of

◀ Irène and Frédéric Joliot-Curie were awarded the 1934 Nobel prize for chemistry for their joint work on artificial radioactivity and the pair played a central role in the French research community of the late 1930s

▼ Traditionally seen as the gateway to the USA, New York – here displaying its famous Manhattan skyscrapers – has all the characteristics needed of a global metropolis

the Wall Street Crash on 'Black Thursday' in October 1929, which led to the Great Depression. And its main island gave its name to the Manhattan Project, which united scientists from around the world in devising the most lethal invention ever: the atom bomb.

THE STORY OF INVENTIONS

Liquid-propellant rockets and television were both invented in the 1920s and both have had enormous influence on the shape of the modern world. Rockets were conceived with space exploration in mind, but were soon put to military use in the Second World War and the subsequent race to develop more powerful missiles. Television's progress was halted by the war, but later it would prove unstoppable. The two inventions came together to make history in the televised Moon landing of July 1969.

From dreams of space to lethal weapons

The people behind the development of modern rocketry were dreamers passionately concerned with space research, but their inventions were quickly put to military use as missiles. After 1945 the descendants of the German V-2s would serve both as strategic weapons and as vehicles of exploration.

On 16 March, 1926, Robert Hutchings Goddard's neighbours were once more having to put up with the whistles and whooshes that his inventions regularly emitted. That day, on a farm outside Auburn, Massachusetts, a small rocket linked by two metal tubes to a fuel tank flew briefly before landing in a cabbage patch. Goddard had just carried out the first successful launch of a liquid-propellant rocket. The time had come at last when he could put his ideas into practice.

Rocket man
Robert Goddard stands alongside the prototype of the first liquid-propellant rocket, which would travel a distance of 12.5m (40ft) in 2.5 seconds.

Futuristic vision
Explorers stand on the lunar surface in a scene from Fritz Lang's The Woman in the Moon *(1929). The rocketry pioneers Hermann Oberth and Fritz von Opel served as consultants for the film.*

The fact was that theoretical reflections on the possibility of space flight had flourished since the turn of the century, with the Russian Konstantin Tsiolkovsky leading the way, followed not just by Goddard himself but also by fellow-Russian Yuri Kondratyuk, Hermann Oberth of Germany and Robert Esnault-Pelterie of France. The theorists were generally of the opinion that only a giant rocket could lift a capsule containing space explorers out of the Earth's gravitational field. Their ideas were ahead of their time – they were already envisaging multi-stage rockets, orbital space stations and landing modules. But the gunpowder-driven rockets then available, including those used in warfare, were barely more powerful than fireworks.

Liquid or solid fuel?

It was already apparent that gunpowder would not suffice to get humans into space. But what should replace it? Rocket propulsion relies on setting off a chemical reaction between a combustible substance and an oxidising agent, represented in gunpowder's case on the one hand by charcoal and sulphur and on the other by saltpetre. Combustion releases gases that are directed backward through a pipe, driving the rocket forward on the action–reaction principle. It followed that there was a crucial need to find good reactive agents to serve as propellants.

arrangements of tubes to feed in the different constituents and turbopumps to pressurise them. For all these reasons, Goddard's 1926 achievement immediately caught the attention of his fellow-researchers, many of whom were already working on similar designs, seeking to find an ideal propellant that was at one and the same time light, easy to store and not too dangerous to handle.

Differing approaches

In Germany in 1923 Oberth had published *By Rocket into Interplanetary Space*, which had aroused widespread interest. Enthusiasts gathered in the Berlin region to try out their inventions. On 5 July, 1927, Oberth and two associates, Johannes Winkler and Max Valier, created the *Verein für Raumschiffsfahrt* (the VfR, or 'Spaceflight Society'). Shortly after, Fritz Lang's silent film *The Woman in the*

The next question was whether to go with solid fuels, like the traditional gunpowder, or with liquid propellants. One advantage of solid-fuel motors was their simplicity: all that was required was to fill the body of the rocket with the mixture, leaving a cavity in the centre as a combustion chamber. The downside was that they only functioned very briefly and could not be extinguished once lit. For interplanetary flight, which requires a long phase of initial thrust plus the later option of switching the motor on and off, liquid fuels seemed the only possible choice. Using them, though, required a sophisticated delivery system featuring isolated fuel tanks, complex

GOING IT ALONE

In 1914 Robert Goddard took out two crucial patents. One involved multi-stage rockets, the other two specific liquid propellants: gasoline and nitrous oxide. The next year he began to experiment with rocketry using his own funds, later supplemented by grants from the Smithsonian Institute and from Clark University, where he taught. For the US army he designed the bazooka, which was first tested just five days before the Armistice of 1918. A year later he published *A Method of Reaching Extreme Altitudes*, a work that would influence a generation of rocket engineers. At the time many were critical – the *New York Times* mocked his ideas in an editorial – and Goddard continued his work in isolation. The aviator Charles Lindbergh was one of his rare backers. In 1930 Goddard moved to New Mexico, where he built rockets of increasing power that were stabilised by gyroscopes; he also worked on cooling systems and turbopumps, but without great success. The US military only came round to his way of thinking after his death in 1945.

THE SAD STORY OF ROCKET-PROPELLED CARS

In 1928 Fritz von Opel, the son of the automobile magnate, decided to apply rocket propulsion to cars and a glider. He approached Friedrich Sander, a fireworks manufacturer, and Max Valier, a co-founder of the *Verein für Raumschiffsfahrt* (see above). That March a first experimental vehicle, equipped with a dozen rockets, attained a speed of 110km/h (68mph). A month later Opel himself reached 200km/h (125mph) in a 24-rocket model. Next year the vehicle was adapted to travel on rails and touched 250km/h (155mph). In 1930 Valier, working with a colleague named Walter Rieder, sought to develop a liquid-fuelled steel propulsion unit, but the rocket exploded in the laboratory, killing Valier outright. Rieder went on to construct a new vehicle in 1931 incorporating thrusters designed by Alfons Pietsch. But that proved the end of the road: rockets, which could only burn for minutes at the most, held little long-term interest for automobile designers.

In the USSR Tsiolkovsky inspired a generation of students, including Friedrich Zander and Sergei Korolev. In December 1931 they founded in Moscow the Group for the Study of Jet Propulsion, generally known as GIRD from its Russian acronym; similar groups were set up in other cities across the nation. Korolev was elected overall director in May 1932. The Nazi takeover of power in neighbouring Germany the following year darkened the international climate, and the Soviet army soon took charge of the venture. Renamed the Jet Propulsion Research Institute (RNII), its Moscow branch was commanded by a military engineer and the focus of its attention was on cruise missiles, jet airplanes and launch rockets. GIRD-09 and GIRD-10 rockets were successfully launched in 1933.

The coming of missiles

By that time it was becoming clear that a new conflict was looming. It was also evident that the heavy artillery used in the past was becoming outmoded in an age of aeroplanes and tank warfare. The time had come for the development of missiles provided with their own thrust engines, stabilising systems and exploding nose-cones. When war did break out,

Moon did much to popularise the idea of space exploration. Meanwhile the VfR was busy developing the Mirak I, II and III rockets. By 1929 the German authorities were showing an interest, seeing in the missiles a way of getting round the provisions of the Treaty of Versailles that prevented Germany from making artillery weapons with a range of more than 35km (22 miles). The VfR's research findings, which until then had remained secret, were handed over to a specially appointed government bureau, clearly indicating that control of the operation had shifted out of the hands of space enthusiasts and into those of the military.

In 1930 a test range was set up on army land at Kummersdorf, where on 14 March, 1931, Winkler launched the first successful German liquid-propellant rocket. Wernher von Braun, then a 20-year-old student engineer in Berlin, arrived in 1932 and supervised the development of a series of Aggregat rockets. One of these, the A-2, reached an altitude of 2,200m (7,200ft) in 1934.

Rocket pioneers
Members of the German VfR rocket society prepare for an experimental launch in 1932 (above). Their work inspired Wernher von Braun's thesis on liquid-propellant missiles, presented to the University of Berlin in 1934.

Russian rocket
The pioneering GIRD-07 (right) was designed in about 1930. It was later adapted to make the Soviet army's first liquid-propellant missiles.

Rocket launcher
Photographed in 1937, this track was used by the RNII to launch the first Russian liquid-propulsion rockets.

THE MAN WITHOUT A NAME

Sergei Petrovich Korolev (1907–1966, above) entered the Moscow State Technical University in 1926. Having read Tsiolkovsky's writings, he became one of the founders of the GIRD rocketry group and subsequently became deputy head of its successor, the RNII. In 1938 he was arrested during Stalin's Great Purge and was tortured and imprisoned. In prison he met the aircraft designer Andrei Tupolev, who was directing a research faculty in the gulag where captive scientists were permitted to work, under lock and key, for the military. Together the two men developed rocket-planes as well as Katyusha missiles. Released on parole in July 1944, Korolev was sent to Germany to assess the V-2 programme. In August 1946, he was appointed chief of a new research bureau, the NII, where he supervised the development of the R-1 rocket, the Russian equivalent of the V-2, and later the nation's entire ballistic programme. He also used political and economic arguments to persuade the state to invest in space research. Yet even though he was the inspiration behind the Zemiorka, Vostok and Soyuz programmes, Korolev's name was kept a state secret and was not publicly revealed until after his death.

Katyushas in action *Lorry-borne Katyusha launchers firing rockets during the Battle of Stalingrad in 1942. The missiles were powered by a nitrocellulose-based solid-fuel propellant and had cross-shaped fins of pressed steel to keep them stable in flight.*

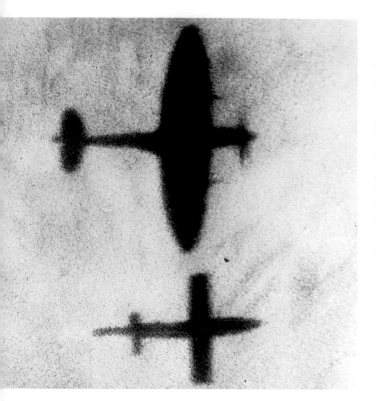

Deadly pursuit
Seen in silhouette, a British fighter plane seeks to deflect an incoming V-1 missile. RAF pilots aimed to bring their wing-tips as close as possible to the wing of the flying bomb, while avoiding direct contact. The pressure of compressed air between the craft could be enough to push the missile off target.

German and Soviet troops at the Front were soon both making heavy use of small solid-fuel rockets with a range of a few kilometres. Katysuha rocket-launchers, known as Guards Mortars or 'Stalin's organs', lined up against the German Nebelwerfer and Panzerwerfer.

More significantly, the Second World War saw the development of new generations of missiles capable of delivering substantial warheads over long distances. At Peenemünde in northern Germany, Von Braun's group was developing the V-1, the first operational cruise missile, and subsequently the V-2 weapon. The V-1 was a flying bomb equipped with stabilising wings and a pulse-jet engine. From June 1944, thousands of V-1 missiles fell on London, Antwerp and Liège, although the onslaught did not go unchallenged: more than 2,000 were shot down or knocked off course by RAF fighter pilots. The V-2, launched in September 1944, was the world's first ballistic missile. Fortunately, it arrived too late to change the course of the war.

GUIDANCE SYSTEMS

Much thought has gone into providing missiles with reliable means of hitting their intended target. On the battlefield they could by guided by radio or through long, fine wires connected to the weapon. These early methods gave way to remote-control systems employing lasers or radar. Another guidance approach was to let the missile itself find the target, for example by detecting the infrared rays given off by the heat of an aeroplane engine. Cruise missiles flying at low altitude are guided on predetermined trajectories by on-board radar. Strategic missiles have gyroscopic stabilisation systems linked to GPS positioning.

Ready to fire
The V-1 could be launched from the ground (right), with its pulse-jet engine already firing, or it could be fired from an aeroplane.

The V-2's post-war heritage

In the 1950s Germany's rocket-planes opened the way to different types of tactical missiles: ground-to-ground missiles, ground-to-air, air-to-ground, air-to-air and sea-to-sea. The designations indicate where the craft is fired from and the type of target: a ground-to-air missile, for example, is one fired from the ground at an aircraft. These short-to-medium-range projectiles were designed to hit enemy positions, provide anti-aircraft defence or destroy tanks or battleships.

Yet this was all fairly small-scale stuff in comparison with the strategic revolution that was getting under way. Atomic weaponry had become a reality for the USA in 1945 and four years later for the Soviets. Only very long-range bombers or ballistic missiles, fired from the ground or from submarines, were suitable for delivering atomic warheads between continents; of the two, ballistic missiles were much less expensive to operate. The former allies of the Second World War now found themselves confronting one another in the long-running Cold War based on the principle of nuclear deterrence.

In a short time ballistic missiles, which were rapid, self-contained and difficult to detect, became the main instruments of deterrence for both camps. An arms race soon set in. Bigger and more powerful rockets with increasing numbers of stages succeeded one another. By the mid 1950s their range was already in the thousands of kilometres; by the 1970s they could travel across inner space to reach any point on Earth.

Many strategists at the time reached the conclusion that missiles were the weapons of the future and that aeroplanes were outmoded, but this view would proved mistaken. For every offensive development, a defensive counter-measure was soon proposed. Improved methods of detection, radar jamming and anti-ballistic missiles were all tried, culminating in the Strategic Defence Initiative, launched in 1983 by US President Ronald Reagan. If implemented, this would have deployed a mixture of missiles, laser weapons and surveillance satellites to shoot down incoming weaponry.

In partial payback to the rocket pioneers, the search for ever-more-powerful delivery systems for nuclear warheads also helped to advance the cause of space exploration. The very first intercontinental ballistic missile, for example – the Russian R-7 Semyorka – also launched the Sputnik satellites in 1957.

Test flight
The first V-2 to test successfully rises from the Peenemünde rocket base on the Usedom peninsula in northern Germany (right). The chequerboard design helped engineers to analyse the rocket's movements when they replayed film taken of the launch.

NAZI ARSENAL

The V-2 was not the only rocket-powered missile developed by the Germans in the Second World War. The Henschel Hs 293A flying bomb, the Kramer X-5 air-to-air missile, the V-4 short-range multi-stage rocket and the A-5 Wasserfall, a miniature V-2, were designed for anti-aircraft defence. The lack of accurate guidance systems meant that none proved decisive.

A FEARSOME WEAPON

Initially known as the A-4, the V-2 was renamed by Hitler with the V standing for *Vergeltungswaffe* – 'vengeance weapon'. They were known to Britons as doodlebugs or buzz bombs from the distinctive sound they made. Developed by Wernher von Braun, it was a liquid-propellant rocket using ethanol for fuel with oxygen as the combustion agent. It stood 14m (46ft) high, weighed 12.5 tonnes and could carry 750kg of explosives for 320km (200 miles). At the peak of its trajectory it reached the outer edge of the Earth's atmosphere 100km (60 miles) above the planet's surface. The first successful trial took place on 3 October, 1942. The missile was first deployed on 8 September, 1944, when it was fired from Belgium on Paris. Technically the V-2 is the forerunner of all rockets, whether used for space exploration or as weapons.

Aerosol sprays 1926

Personal grooming
A can-wielding chorus accompany a song in the 2007 musical Hairspray.

The Norwegian chemist Erik Rotheim got the idea for the modern spray can from old-fashioned perfume dispensers with bulb sprays. Casting around for a more efficient mechanism than manual squeezing, he came up with a canister enclosing a gas maintained under pressure in a liquid state; when a valve was released, the gas and liquid were jointly ejected in the form of an aerosol spray. In 1926 Rotheim took out a patent on his invention, which he tried to exploit commercially without success. Eventually he put the patent up for sale, but the concept languished until the 1940s, when the US government was looking for an efficient way of protecting its troops serving in the Pacific from malaria-carrying mosquitoes. In 1941 two researchers, Lyle Goodhue and William Sullivan, devised a 'bug bomb' that proved popular with the soldiers, as it was easy to use and carry. They patented the device two years later.

The American inventor Robert H Abplanalp improved on the design in 1949 by developing a light, cheap can fitted with a fixed valve that made it easy to control the pressure of the spray. Aerosols quickly became phenomenally successful, being used to package mass-market products as diverse as insecticides, cosmetics, paint, air fresheners, shaving foam and whipped cream.

CHLOROFLUOROCARBONS

The propellants used in the first aerosol sprays included ammonia, chloromethane and sulphur dioxide. They all had their downsides, being toxic, inflammable or potentially explosive. They were rapidly replaced by chlorofluorocarbons (CFCs) after the American chemist Thomas Midgley found an improved way of synthesising the compounds in 1928. Odourless, cheap to make, barely flammable and extremely stable, CFCs were seized upon for many industrial uses, but in the 1970s scientists established that they, too, were highly dangerous: they were destroying the protective ozone layer in the Earth's atmosphere. The Montreal Protocol, signed by 24 nations in 1987, ordered the phasing out CFCs to be replaced by less harmful alternatives, such as butane-propane mixtures, nitrous oxide and carbon dioxide.

OZONE KILLER

The steam iron 1926

The first alternatives to ancient flat irons and box irons filled with hot coals were cumbersome implements heated by liquid fuels like kerosene or gas. The coming of electricity brought new possibilities. An American, Henry Seely, was responsible for the first electric iron, invented in 1882. Some early models featured a carbon arc, with the electrical discharge passing through air; when these proved too dangerous, a more resistant design was introduced in 1892. The first electric steam iron, in which water was heated to aid the ironing process, appeared in 1926, when a New York dry-cleaning firm, Eldec, started making devices it sold on to other industrial users. A controllable thermostat followed soon after. Steam irons became standard domestic appliances in the 1950s, since when improvements include steam controls, spray functions and wire-free designs.

Household help
Ironing in the 1950s (above), the decade in which Hoover introduced the steam iron to the UK.

Front-wheel drive 1926

Cars are generally propelled forward by the rotation of the rear wheels. The front wheels improve road-holding, but getting them to transmit the driving force generated by the motor without jolting was always going to be difficult. Two Frenchmen, Jean Albert Grégoire and Pierre Fenaille, addressed the challenge in 1926 by equipping a prototype vehicle with a robust universal joint. They went on to prove the worth of their invention in the 24-hour race at Le Mans the following year. On the strength of this success they exploited their patent in Europe, the USA and Russia, and founded the Tracta firm, which manufactured cars under its own name until 1933. Mass production followed in 1934 when Citroën introduced the 7A, known in its home country of France as the Citroën Front-Wheel Drive.

Classic car
Manufactured from 1934 to 1957, the Citroën Traction Avant (below) was the car that brought front-wheel drive into mass production. It also featured a pioneering monocoque design, with the car's chassis and bodywork welded together in one piece.

BSA LEAD THE WAY

Britain's first front-wheel-drive vehicles were three-wheelers built by BSA. The acronym stands for Birmingham Small Arms Company – the firm was originally set up by gunsmiths in 1861, then ventured into bicycles in 1880. The first motorcycle came in 1905 and a prototype car two years later. The three-wheel design, with two wheels at the front, was already associated with Morgan when BSA took it up in 1929, opting for front-wheel drive to reduce the stress on the single rear wheel. Two years later the firm also introduced front-wheel-drive four-wheeled cars and vans. The last models were made in 1936.

TELEVISION – 1926

The small screen arrives on the scene

Initially seen as having no practical purpose, television would come to occupy a central position in almost every household. Although its influence has been much decried, the new medium brought knowledge and culture within everyone's reach.

Improvised invention
The televisor constructed by Baird in 1925 had a Nipkow disc made out of a hatbox. The motor that turned it rested on wood from a tea chest.

THE NIPKOW DISC

In 1884 a German student named Paul Nipkow devised the first system for breaking down an image for electrical transmission. It took the form of a perforated disc, with the number of holes corresponding to the number of lines used to reconstitute the image, from an initial 18 up to 200. The holes were arranged in a spiral, and when the disc started to turn the first of them captured the top left-hand fragment of the image, following it in a curving line across its width. The second hole continued the scanning process, beginning marginally lower down; a third followed, and so on to the end. The light passing through the holes activated selenium photocells at the rear of the disc, making it possible to convert the image into an electric signal.

Pioneer performance
Scottish engineer John Logie Baird prepares an experimental transmission in 1928 (above). The system that Baird devised called for such bright illumination that initially he preferred to work with heads from ventriloquists' dummies rather than people.

One October day in 1925, John Logie Baird emerged from his Soho workshop to stop the first person who crossed his path in the street. This was how a 20-year-old office worker named William Taynton got to play a part in a revolutionary experiment. He found himself seated in front of a revolving wooden disc about 60cm (2ft) wide, pierced by 30 holes. Baird disappeared into a neighbouring room, where he feverishly scrutinised a small screen. Nothing showed up. He returned to the studio to find his unwilling assistant cowering in a corner, scared by the intensity of the light required for the trial. Baird did his best to reassure him, offering him money to stay put in front of the disc. This time Baird saw an image on the screen, somewhere between black and white and sepia in colour. Though it was blurred and slightly distorted, he recognised the nervous features of his helpmate, who had no idea that he had just become the first person in history to appear on television.

Electromechanical beginnings

On 26 January, 1926, Baird re-created the experiment for members of the Royal Institution and a reporter from *The Times*. He presented them with a 30-line moving image of a head, scanned five times a second. His apparatus consisted of a big wooden

THE VIDEOPHONE

On 7 April, 1927, Herbert Hoover – then US Secretary of State for Commerce – took a long-distance telephone call from Walter Sherman Gifford, president of the AT&T telephone company. What made the call special was that Gifford could see Hoover's face on a screen in a demonstration of the world's first videophone link, between New York and Washington DC some 200 miles away. Communication was one-way only; Hoover could not see Gifford. The system used the Nipkow disc and three telephone lines, one of which transmitted the image and a second the voices, while the third sent a signal that synchronised the two. The screen measured just 6cm^2, diffusing sixteen 50-line images a second.

Historic image
The oldest surviving photograph of a televised image, dating from about 1926, shows the head of John Logie Baird's business partner Oliver Hutchinson.

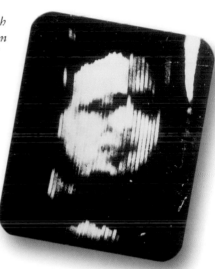

Live demonstration
Baird set up the world's first TV studio in 1929 and travelled widely to demonstrate his system. He is seen here with the German scientific writer Dr Alfred Gradenwitz (seated, below).

Nipkow disc (see box, bottom left) and a system of lenses designed to concentrate light on photosensitive cells. The Nipkow disc broke the image down into electrical impulses, whose varying intensity was translated into areas of shadow or brightness at the time of transmission, when a variable light source reconstituted the impulses into differing degrees of light and shade. A second Nipkow disc sharpened the end result, which was made up of points of light of varying intensity. The human brain duly reconstructs these into solid images thanks to the phenomenon known as persistence of vision.

At first Baird's researches were met with at best amusement and at worst disdain. Television – the word had been coined in 1900 by the Russian physicist Constantin Perskyi at the Paris World's Fair – seemed of interest only to a handful of technology buffs. An article in the journal *Scientific American* went so far as to state that, even if the transmission of images was technically feasible, it was never likely to have any practical applications.

Yet Baird refused to give up. He worked not just on the development of a mass-market television set but also on long-distance signal transmission, which involved increasing the power of the transmitter and the sensitivity of the receiver. In 1927 he broadcast a moving image of a face from London to Glasgow. A year later he established a transatlantic

Improved design
EMI developed the Super-Emitron video tube for the BBC (left). Inspired by the iconoscope, the device supplanted Baird's camera in 1937.

No moving parts
American inventor Philo Farnsworth in 1929 with his Image Dissector, the first all-electronic television system with no mechanical parts.

television link between Britain and New York State. In 1929 the BBC, which had been founded for radio in 1922, began to broadcast programmes made by Baird's company for the benefit of owners of 'televisors', also produced by Baird's firm and first marketed in that year. By 1932 almost 3,000 Britons owned sets, which displayed small, jerky images. The service was on air for just a few hours a day, if at all, mostly featuring well-known singers or actors performing. As the televisor set had no loud-speaker of its own, the sound was provided through a radio set placed beside it.

The all-electronic alternative

Baird's invention seemed to have a great future ahead of it, but technologically it was already being overtaken by another system, the brainchild of an Idaho farmer's son. In 1921 the 15-year-old Philo Farnsworth was digging potatoes on his Mormon parents' smallholding. But he was not destined for a life on the land. He proved a brilliant pupil at school, showing a passion for electricity and wireless transmission. Legend has it that he got the idea for an electronically scanned television system by staring at the parallel furrows lining the fields where he worked. Yet even then the concept was not entirely new, for as early as 1911 a Scottish electrical engineer named Alan Campbell-Swinton had already outlined a way of using a cathode-ray tube to transmit images over distance.

In 1927 Farnsworth became the first person to put Campbell-Swinton's ideas into practice. He soon faced fierce competition from a US-based Russian émigré, Vladimir Zworykin, who had developed a purely electronic scanner that he called the iconoscope, generally considered to have been the first true television camera. Zworykin had been working on the idea since 1923, but it was the early 1930s before practical versions of the apparatus became available, by which time Farnsworth was accusing the Russian of plagiarising his invention. In 1934 the US Patent Office decided in Farnsworth's favour, giving him the right to consider himself the father of all-electronic television. Yet it was Zworykin's iconoscope, which had greater light sensitivity than Farnsworth's tube, that became the standard

Transatlantic transmission
On 9 February, 1928, onlookers gather in Coulsdon, Surrey, to witness the transmission of the first live images across the Atlantic to Hartsdale in New York State. The transmission was made using John Logie Baird's electromechanical television system.

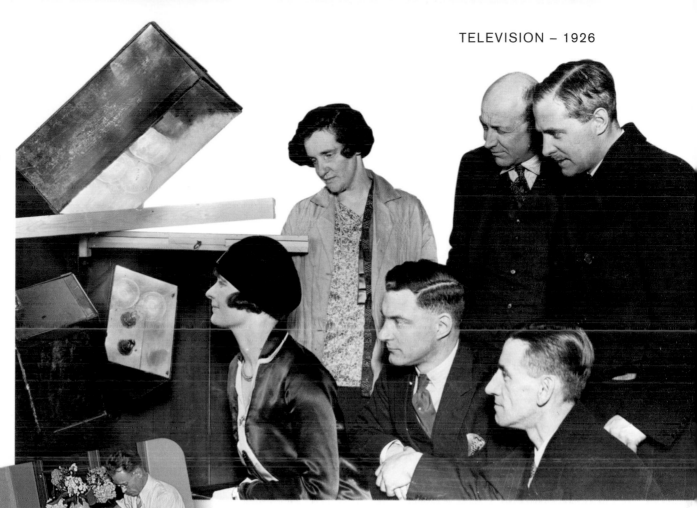

Sound and vision
Philo Farnsworth (left) demonstrates a television set to journalists (unseen in the photograph). The set could receive both radio and television broadcasts.

Image maker
Vladimir Zworykin (below) in 1929 with the cathode-ray tube that he invented.

camera across much of the world. In 1933 RCA, the US corporation for which Zworykin worked, started production-line manufacture of iconoscope machines, which could capture 120-line images 24 times a second.

Improving standards

By the early 1930s manufacturers had worked out how to provide the vacuum needed within the cathode-ray tubes that lay behind the fluorescent screens. At roughly the same time, interlaced electronic scanning came into use: the bundle of electrons in the tube scanned alternate lines in succession, thereby reducing

THE TELEVISION SCREEN

In 1897 the German physicist Karl Ferdinand Braun invented the oscilloscope, an instrument that could display high-frequency electric currents. It consisted of a cathode-ray tube, in which a flow of electrons (known at the time as cathode rays) traversed a partial vacuum to strike a fluorescent screen. Through the use of electromagnets it was possible to deflect the beam and so draw a line on a screen, whose phosphorescence made the trace left by the electrons visible. Through the phenomenon known as 'persistence of vision', what was actually a point of light in motion appeared as a continuous line. In 1907 a Russian scientist called Boris Rosing had the idea of using the Braun tube to reconstitute images. The first television sets using cathode-ray technology went on sale in 1933.

Live TV
A cameraman from the Farnsworth Television Laboratories in Philadelphia prepares to film a local band in July 1935.

jerkiness in the televised image. In 1935 EMI introduced the first 405-line cathode-ray sets, marking the victory of electronic television over its electromechanical rival. At first the BBC used both systems, but it was obvious that the all-electronic version was superior and in 1937 the Corporation opted exclusively for the 405-line standard. In the same year France introduced a 455-line electronic system. Over the next two years, up to the outbreak of the war, the number of viewers in Britain increased tenfold from just 2,000 to 20,000.

The Second World War called a halt to the medium's development in Europe. Once peace had been restored, television broadcasting began again with each nation adopting its own chosen standard: 405 lines in Britain, 441 in Germany and in Italy, 455 in France. In 1952 a higher-resolution 625-line model was adopted across much of continental Europe, with Britain eventually following suit in 1964, when BBC2 went on air. Even then, the 405-line VHF transmissions needed for compatibility with older sets continued for two more decades in Britain, only finally being dropped in 1985.

Spectator sport
German viewers crowd around a television screen in 1952 to watch a football match (below). The game was broadcast by NWDR, successor to Radio Hamburg, set up by the British after the war.

TELEVISION FIRSTS

In 1927 a comedian by the name of Dolan became the first television presenter when he fronted an experimental broadcast produced by the AT&T corporation. In 1931 a presenter called Stein became the first talk-show host on a show produced by the experimental W2XCD channel in Passaic, New Jersey. That same year John Logie Baird televised the first live tranmission of the Epsom Derby. Three years later a German woman called Patzschke became the first continuity presenter. In 1938 the BBC introduced 'New Map', the first current-affairs magazine programme, and 'Spelling Bee', the first game show. NBC and CBS were the earliest commercial stations, licensed in the USA in 1941.

COLOUR ON THE SCREEN

John Logie Baird experimented with colour television in 1928, using three Nipkow discs equipped respectively with red, green and blue filters. In 1940 CBS broadcast in colour in the USA employing a system, developed by Peter Goodmark, that used rotating discs on both the black-and-white cameras used for transmission and on the sets that received the images. Ten years later this arrangement became the national standard and in 1951 CBS began to manufacture colour sets. But soon afterward the authorities changed tack, backing a rival system developed by RCA, which in turn became the NTSC (National Television System Committee) standard. Superimposing three images, respectively red, green and blue, on screen to provide colour, this became the regular analogue model across much of the world. The first sets came on the market in 1954. Colour television reached Britain in 1967, at first only on BBC2; BBC1 and ITV began colour broadcasts two years later. Digital TV began to replace analogue in the USA in 1997 and in Britain in 1998. The aim is to make it the British national standard by 2012.

Colour on screen
Colour television made its entry in the USA in the 1950s and was spectacularly successful. Between 1947 and 1955, production of TV sets rose from around 178,000 to more than 7 million.

Toward television for all

As channels multiplied, the frequency band quickly became too narrow. Television broadcasts were initially emitted on wavelengths close to those used for radio, but over the course of the decade they migrated to the VHF range, from 30 to 300Mhz, before moving on to ultra-high frequencies above 300MHz in the 1960s. In 1962 the Telstar 1 satellite was launched, making it possible to transmit broadcasts across the Atlantic via space. The small screen had conquered the planet and also helped to shrink it.

In the 1990s flat screens finally did away with the need for cathode-ray tubes, paving the way for high-definition resolution using a 1080-line format. By then television was not just providing information and entertainment. It had found an important role in security surveillance. It had proved to be a valuable tool in medicine, both for monitoring patients and as an essential aid in surgery, revealing internal images from tiny cameras in keyhole operations. Similarly, cameras can be used to relay images from inaccessible places like the heart of nuclear reactors or deepwater drilling rigs. The long-distance transmission of images has become central to the society we live in.

Multi-screen viewing *The proliferation of cable, satellite, ADSL and digital terrestrial television has made it possible for TV viewers to access literally thousands of channels from all around the world.*

Television

Inventions rarely arrive in their final form Rather, they emerge as the fruit of earlier advances that paved the way. By mastering transmission over the airwaves, the inventors of television managed to bring recorded sound and vision into people's homes.

CAPTURING IMAGES
BIRTH OF THE TV CAMERA

In 1817 Swedish chemist Jons Jacob Berzelius discovered that selenium is light-sensitive. In 1873 Britons Willoughby Smith and Joseph May confirmed his findings, establishing that the element's resistance to electric currents changed when it was exposed to light. Building on this foundation, in 1923 Vladimir Zworykin invented the iconoscope, ancestor of the modern TV camera, using the sensitivity of silver to light to convert images into electrical impulses in the form of video signals. EMI improved the design in 1935 with the more sensitive and compact Super-Emitron tube. Improved versions of the Super-Emitron were used in cameras up until the mid 1980s, when they were replaced by digital equipment.

Vinten H 35mm camera designed by John Logie Baird in 1936

TV transmitter mast

RECORDING SOUND
FROM THE EAR TRUMPET TO THE MICROPHONE

Working independently, Thomas Edison and Émile Berliner both came up with the idea of the microphone in the same year – 1877. In early recordings for Berliner's 'gramophone', the sound source was directed toward an ear trumpet that channelled the sound waves to a needle which in turn etched a continuous groove on a wax disc. The second generation of microphones used an amplifier to transform sound waves into electrical signals. Condenser microphones, developed by Bell Laboratories in 1916, considerably improved the quality of recordings by reducing background noise; they would be used to produce the first 'talkies' from 1927 on. In recent years miniaturisation and other advances in electronics have made microphones smaller and more efficient.

Filming a broadcast in 1936 using the EMI all-electronic system

Marconi ribbon microphone used by the BBC from 1934 to 1959

TRANSMITTING SOUND AND IMAGES
FROM WAVES TO CURRENTS

In 1887 Heinrich Hertz demonstrated that ultra-violet rays cause certain metals to give off electricity. Three years later Édouard Branly invented the 'coherer' – an insulated tube containing iron filings, linked to a battery and a galvanometer – which provided a practical means of detecting electromagnetic waves. In 1895 Marconi managed to send a message in Morse code over a distance of a mile via electromagnetic waves transmitted and received through antennae. In 1900 Reginald Fessenden succeeded in transmitting the human voice using a high-frequency spark device to modulate the same wireless waves that in the 1930s would be used to carry video signals for early television broadcasts. In 1904 Ambrose Fleming invented the vacuum tube or diode, which turned out to be an improvement on Branly's coherer as a receiver. Diodes became triodes in 1906 thanks to contributions by Lee De Forest, improving sensitivity and amplifying the signal. Television sets were equipped with triodes until the 1950s, when they were replaced by transistors.

Outdoor TV aerial

CATHODE-RAY TUBES
FROM OSCILLOSCOPE TO SCREEN

In 1875, while studying electrical conductivity of gases at low pressure, the physicist William Crookes discovered that cathodes emit rays that cause luminous marks to appear on a fluorescent screen. He duly named them cathode rays. Twenty-two years later, in 1897, Joseph John Thomson was able to show that the 'rays' were in fact streams of electrons. That same year Karl Ferdinand Braun used an improved Crookes tube to study electric currents and to develop the oscilloscope. A decade later Boris Rosing had the idea of using Braun's cathode-ray tube as a receiver.

A Ferranti TV receiver of 1930

Joint radio and TV set of 1938

THE FUTURE OF TELEVISION
FROM ANALOGUE TO DIGITAL

In modern televisions bulky cathode-ray tubes have already been replaced by flat-screen technology. Television is going digital and achieving high definition, and may now be retransmitted by cable or satellite or in digital terrestrial form. Screens are rapidly being miniaturised and are becoming increasingly portable. Television services can also be accessed via the internet and mobile phones.

LOUDSPEAKERS
FROM RODS TO MEMBRANES

In 1785 Pierre-Simon Laplace discovered that a metal rod conducting an electric current starts to move when it enters a magnetic field. A century later, in 1877, Wernher von Siemens made use of this 'Laplace force' to convert the varying intensity of electric currents into vibrations in a membrane, producing sound. At first the weakness of the currents deployed limited the application of his work to radio and telephone transmissions. But with the coming of triodes the signal could be amplified to provide vastly improved sound quality, an advance from which the first TV sets benefited. Some sets were fitted with integral loud-speakers from the start.

A modern loudspeaker

Prototype miniature TV screen made of plastic

The strange new world of particles

Following on from Einstein's theory of relativity, the second great revolution in modern physics, which had been germinating since the turn of the century, came to fruition in 1926. The new quantum mechanics described the behaviour of elementary particles, marking the point where science finally parted company with everyday common sense.

Meeting of minds
Two of the founding fathers of quantum mechanics, Paul Dirac (on the left) and Richard Feynman, deep in conversation at an international conference in Warsaw in 1962.

Between January and June 1926 Erwin Schrödinger, a professor of theoretical physics at the University of Zurich in Switzerland, published four articles in the German scientific journal *Annalen der Physik* (the 'Annals of Physics'). Between them, these articles set out a completely new theory at the very heart of science: quantum mechanics, the ultimate explanation of the interactions between particles, atoms and molecules, excepting the gravitational relationships explained by relativity. The new proposals not only marked a breakthrough in the realm of ideas but also brought practical results, for they would lead to important inventions, semi-conductors and lasers among them.

PLANCK'S BLACK-BODY RADIATION

Max Planck's black-body experiment described what happens when an object – say, a chicken – is placed in an air-free oven that has been pre-heated by electromagnetic waves in the form of infrared radiation. At first the chicken absorbs the waves. Then after a time, as the chicken heats up, it begins to radiate energy itself up to the point at which oven and chicken are at the same temperature. By this time it is giving off as much heat as it absorbs. Unless quantum theory is applied, the energy-exchange equation involved ends up with an infinite number.

The universe's diffuse cosmological background
The tiny fraction of heat radiation left over from the Big Bang has black-body characteristics. The amount of energy remains vast, but we are not bombarded by it because it is made up of quanta maintained at a precise, calculable level.

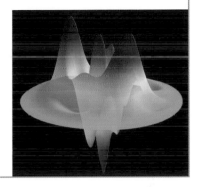

The significance of this super-theory to modern science can hardly be overstated. All physicists today learn about quantum mechancics from their first year at university. And yet, as the Nobel prize-winning American physicist Richard Feynman ironically liked to point out, it is a theory that 'nobody really understands'.

Making no sense

Feynman was hardly exaggerating, for if quantum theory is to be believed there is a level of the infinitely small at which space, time, energy and matter make discontinuous leaps and particles display a quality of ubiquity that enables them to occupy different locations or separate physical states at the same time. In this they resemble Schrödinger's famous cat (see box above), which was simultaneously here and not here, dead and yet alive. Even more extraordinarily, the theory maintains that two particles can form a single unity even when separated by a distance of several miles. Standard concepts of reality – such as insisting, for example, that a cat must be either dead or alive but not both at the same time – become lost in such riddles. The proponents of quantum theory would answer that the cat analogy is merely an approximation, transporting ideas that belong in the microcosmic world of particles to our own macrocosmic level. But even so, if quantum theory had not been confirmed experimentally many times over, it would be hard to expect anyone to credit it, so bizarre can it seem, not in its equations but in the way it describes the world.

Planck's oven

Science may have made its great leap into weirdness in 1926, but the jump had been prepared as early as 1900, when Max Planck first raised the problem of black-body radiation (see box, bottom left). At the time the German physicist was having conceptual trouble with ovens. He found he could not fully explain the heat exchange (in terms of thermal energy) between the oven itself and objects put in it to heat up; his theoretical calculations systematically led him to the aberrant conclusion that the amount of energy exchanged must be infinite.

Planck came up with the proposition that the energy absorbed and radiated by the object in the oven takes the form of packets – rather as if one were to visualise the flow of a river not as continuous but as discrete, made up of minuscule units of water that are separate from one another. By means of this artifice, which involved treating a flow of electromagnetic waves as being composed of indivisible packets of energy, Planck managed to make his

Simultaneous realities

By capturing light in a photon box for a sufficient length of time, as shown in this photographic simulation (above), it is possible to reveal two simultaneous but different states, the one reflecting classic (red) and the other quantum (blue) physics. The duality is a reflection of Schrödinger's paradox of the cat that is at the same time dead and alive.

Circle and square

The paradox behind quantum mechanics can be illustrated by an object presenting different characteristics depending on the observer's viewpoint. An electron beam traversing a nickel crystal, represented here as a cylinder, casts a wavelike reflection even though the concentric circles making this up could only be produced by particles (see below). Light thus has wave and particle characteristics at one and the same time.

CONFIRMING THE PARTICLE–WAVE DUALITY

Two Americans, Clinton Davisson and Lester Germer, confirmed the dual nature of electrons with an experiment conducted in 1927. They fired a bundle of electrons from an electron gun at a nickel crystal, obtaining on a fluorescent screen a diffracted pattern of concentric circles alternating with light and dark bands. The fluorescence could only be explained in terms of particles striking the atoms that made up the screen, but the pattern itself, resembling the circles created when a pebble is thrown into water, confirmed Louis de Broglie's view that elementary particles have wave-like properties.

The end of causality? *The logical implication of Aspect's photon experiment (see box, right) is that the same object can be in two different places at the same time, like the frog shown below and far left. The experiment has since been replicated in laboratories around the world, including at the University of Geneva where in 2003 researchers from the Applied Physics Group used an arrangement of lasers (above).*

ASPECT'S EXPERIMENT

In the early 1980s a young French physicist named Alain Aspect conducted for his doctoral thesis a series of experiments that confirmed that photons can continue to be correlated in their wave functions even when separated by a considerable distance. In effect, the wave-like nature of photons allowed him to superpose two of the particles, rather like two waves of water, while their particle aspect enabled him to dispatch them in different directions. He then established that when one of the photons was disturbed, the other was disturbed also, confirming that they remained correlated even over distance. In later experiments scientists have replicated the effect with photons over a distance of up to 30km apart.

calculations work. By doing so he saved physics from absurdity, but only by opening a breach in the classic Newtonian worldview that Einstein was to widen irreversibly.

The particle-wave phenomenon

In 1905 – the same year that he was laying out his theory of relativity – Einstein published an article on the photoelectric effect (seen today in the operation of solar panels), by which atoms of photoelectric material absorb energy from light then emit electrons in the form of an electric current. He proposed that they are able to do so because light is made up of indivisible quanta of the type described by Planck, which act like separate particles that strike the electrons, ejecting them from the atoms.

In effect Einstein had discovered the light particles now known as photons. This was confusing, for people had known for a century or more that light took the form of a wave. With a touch characteristic of his genius, Einstein bridged the apparent paradox by declaring that light was simultaneously a particle and a wave. This notion of particle-wave duality, however alien to

common sense, was to lie at the heart of the future science of quantum physics, even though it would take a couple of decades for such a counter-intuitive idea to catch on.

Matter made of waves

The controversial point was that Einstein's duality principle did not present light as a wave bearing particles in the way that an ocean breaker carries surfers. Rather, it was simultaneously both – a sort of combined surfer-and-wave that could be measured under one or the other aspect but not both at once. In practice, it can be identified in its particle form with the aid of a particle detector – a bubble chamber or calorimeter, for example – and as a wave by a gravitational wave detector, such as a Weber bar or an interferometer, but not simultaneously by the one and the other. The fact is that it is never possible to 'see' a

THE PRACTICAL SIDE OF QUANTUM PHYSICS

Quantum physics can seem remote from daily life, yet many technological innovations have flowed directly from it. First among these are lasers, intense beams of light artificially created by the quantum mechanical process of stimulated emission. Semi-conductors, which are fundamental to modern electronics, also owe their electrical properties to the quantum state of their atomic structure. Quantum technology plays a part, too, in other state-of-the-art devices such as superconductors and scanning tunnelling microscopes (STMs).

particle-wave in its entirety, and for good reason, given that the two manifestations are totally opposed. Waves, in the air or in water, are disturbances with a range of amplitude and frequency, while particles do not vibrate and are, by definition, punctual in the sense in which scientists use the term, meaning confined to a point in space. Combining the two conjoins opposites, like saying that 'wet' is synonymous with 'dry'. So it came as a shock when, in 1924, matter itself was said to be made up of particle-waves.

Constructing a new theory

In that year Louis de Broglie, a French physicist of aristocratic origins, completed a doctoral thesis expanding on Einstein's insight; in de Broglie's words: 'My essential goal was to extend to all particles the coexistence of waves and corpuscles discovered by Einstein in 1905.' In short, he established that the bizarre duality Einstein had tracked down in light existed in all matter. Experiments with electrons, matter's constituent particles, confirmed his theory three years later. The implications were immense. In effect a whole new physics had to be created to account for the behaviour of this bizarre fusion of opposites.

That task was taken up by Schrödinger and Werner Heisenberg. The two physicists, one Austrian and the other German, developed separate theories that both proved prescient. Schrödinger produced an equation describing how the quantum state of physical systems changes over time, while Heisenberg came up with an abstract mathematical model of so-called 'matrix mechanics'. Although the two approaches were very different, they had one thing in common: both were perfectly consistent with the results turned up by experiments conducted on the entities, whether in particle or wave form, and also correctly predicted findings that had yet to be made.

Neither attempted to address the nature of the particle-waves themselves. In fact the two men's work jointly highlighted a distinctive feature of quantum physics in that their theories, unlike those of classical physics, limited themselves to describing rather than explaining what researchers observed. It was a little as though Newton had come up with a theory of gravity that enabled astronomers to correctly predict the Moon's behaviour as long as it was under observation while admitting that at other times, for all they knew, it might be looping the loop.

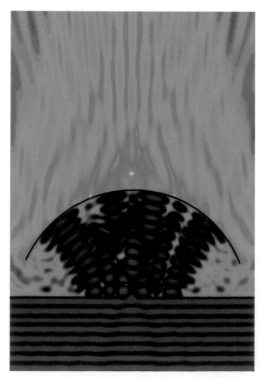

Breakthrough
A beam of electrons, represented by the horizontal lines at the bottom of the image, flows beneath a barrier (the horizontal black line) pierced by a hole. When the electrons strike the barrier, only those whose wavelength exactly coincides with the diameter of the gap penetrate into the cavity above, as defined by the black semi-circle. They accumulate there, vibrating against one another until some diffract beyond the cavity, indicated by the vertical green lines in the top half of the image.

In 1927 the English physicist Paul Dirac formulated an equation to describe the behaviour of electrons at speeds close to that of light, thereby unifying quantum physics and special relativity in what is now called quantum field theory. But he met a problem: the equation demanded the existence of a new kind of 'antiparticle' to transform the existing particles into light energy. A first antiparticle, the positron, was observed experimentally in 1932, providing the earliest evidence of the existence of antimatter.

Photographing atoms *In an experiment carried out in Paris in 2009, caesium atoms trapped by a laser were 'frozen' at 2.5 millionths of a degree above absolute zero. Gravity stopped them escaping, while a laser beam directed through the prism beneath prevented them from falling. From a starting speed of 300m/sec, they were slowed to immobility for a few seconds – long enough to take this photo (left).*

Everywhere and nowhere

The theories of Schrödinger and Heisenberg provided ways of measuring the interactions involved in a quantum system but failed to explain the phenomenon itself, a fact that seemed to run against common sense. Yet it soon became evident, to the great displeasure of Einstein and others, that the impossibility of describing a particle-wave's behaviour other than by measuring it did not necessarily mark a limitation on knowledge. Rather, it flowed from the nature of the quantum system itself. Experiments showed that before quanta are measured, and so forced by the action of the observer to reveal their properties either as waves or particles, they have no precise spatial position (scientists speak of the 'non-locality principle') and no fixed state (a condition known as 'quantum indeterminacy').

An indeterminate world

In 1925 it was still possible to believe that such anomalies were illusory, caused by gaps in the existing knowledge. But on 21 June, 1926, the last of Schrödinger's four articles demonstrated that his wave mechanics and Heisenberg's formalism were equivalent, giving quantum physics a coherent, unified theory. From then on, for all the controversies that followed, a growing consensus would force physicists to admit that the laws governing the particles that underlie all aspects of daily life have a strange indeterminacy written into their very being.

Laser proof *Lasers are among the principal tools of quantum research, enabling scientists to verify and amend the theoretical framework preparing the way for quantum computers.*

The iron lung 1927

Intensive care
A woman paralysed by polio lies in an iron lung at St Luke's Hospital, Chicago, in 1930.

On 12 October, 1928, an unconscious eight-year-old girl patient at the Boston Children's Hospital became the first person in the world to be placed in an iron lung. This life-saving piece of equipment enabled her to breathe even though her respiratory muscles had been paralysed by polio. Properly known as a negative-pressure ventilator, the iron lung had been invented the year before by Philip Drinker of the Harvard Medical School with the help of two physiologists – his own brother Cecil and Louis Agassiz Shaw. The technique involved enclosing the patient's entire body, with the exception of the head, in a steel chamber fitted with a system of pumps that increased and decreased the air pressure causing the lungs to expand and contract, taking in and expelling air.

The devices, which often took up entire hospital rooms, became invaluable, largely because of the prevalence of polio at the time. Thousands of lives were saved, but there were not enough of them. When a major polio epidemic struck Copenhagen in 1952, an anaesthetist named Bjorn Ibsen had the idea of employing a positive-pressure respirator, previously used for anaesthetics, to deliver a continuous flow of air through a mask. The procedure initially involved a tracheotomy (surgical opening of the trachea). Ibsen went on to pioneer intensive-care wards, cutting local mortality from 80 to 25 per cent.

Iron lungs gradually fell out of use in the 1950s as vaccination programmes began to conquer polio. In other cases they were replaced by respirators, which continued to be in demand for patients suffering from chronic or acute respiratory problems. Now increasingly sophisticated in their operation, respirators are still frequently used for cardiopulmonary resuscitation (CPR).

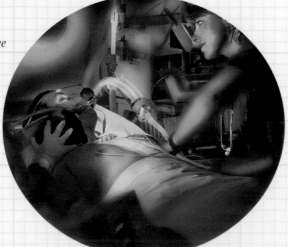

Deep breathing
A patient in intensive care receives air through an oxygen mask. This technique avoids the need to insert a tube in the trachea, reducing the risk of infection and other potential complications.

THE FIRST RESPIRATORS

Scientists turned their minds to artificial respiration in the 17th century. The first realistic respirator, devised in 1832 by Scottish physician John Dalziel, was a tank in which negative pressure was maintained by bellows driven by a piston. In 1881, prompted by the death of his infant son, Alexander Graham Bell devised a 'vacuum jacket'. But the iron lung's true successor was the spirophore, first made in 1876 by French doctor Eugene Woillez to resuscitate babies and victims of drowning.

Crepe soles 1927

In 1927 the crepe rubber sole won for its inventor – a Frenchman named Delbon – the Concours Lépine prize for innovation. Delbon claimed that his new soles were impervious to wear and tear, and over the years since they have proved hard-wearing, although perhaps not as indestructible as their inventor liked to believe. His manufacturing process involved mechanically grinding and shredding latex to form thin leaves that were then left to dry slowly.

Heel to toe
Crepe soles owe their durability to a double thickness of rubber. Some also have a layer of silicon to make them more water-resistant.

Water-resistant watches 1927

In 1926 Hans Wilsdorf, the German founder of the Rolex watch company, took out a patent on an improved way of waterproofing watch cases to protect the inner workings from damage. His device employed a crown winding mechanism that screwed onto a stem riveted to the case. A year later, when 27-year-old Mercedes Gleitze became the first British woman to swim the Channel, she did so with a Rolex watch on her wrist. Wilsdorf used the occasion to put the watch on the market, launching it with a fanfare of publicity in the *Daily Mail*. Since that time other watch manufacturers have come up with alternative ways of keeping air and water out of their cases. The Aquadura system, for example, introduced by Mido in 1934 and still in use, employs a specially treated cork component to seal off the winder opening.

Rolex Oyster
In 1953 Rolex introduced the Submariner (right), the first watch claimed to be watertight down to 100m. The Sea-Dweller 4000 followed in 1980, guaranteed for dives down to 1,220m (4,000ft).

Cross-Channel pioneer
Two weeks after her successful swim across the Channel in October 1927, Mercedes Gleitze (above) tried to repeat the feat, but the cold forced her to give up. A stenographer by profession, she also became the first woman to swim the Straits of Gibraltar (in 1928) and the Dardanelles (1930).

GLOWING DIALS

In the 1920s the luminescent quality of radium, still a recent discovery, was put to use in a variety of ways. There was a vogue for statues painted with radium-enhanced zinc sulfide that glowed in the dark, and cosmetics counters even sold a 'luminescent' lipstick. But the most common use was for watches with yellow-green luminous markings on the dial. These grew in popularity until they were banned in 1945, when the dangers of radioactivity were finally understood. By that time many of the watches had already been consigned to cupboards, as the radium tended to turn the glass face opaque.

A singular achievement

At 10.22pm on 21 May, 1927, after 33 hours and 30 minutes in the air, Charles Lindbergh landed his little monoplane the *Spirit of St Louis* at Paris's Le Bourget aerodrome. Thousands of enthusiastic spectators were there to welcome the 25-year-old American aviator, who had just completed the first solo non-stop flight across the Atlantic.

Lindbergh barely had time to lever himself out of the cockpit and to shout 'Well, I did it!' before his admirers fell on him. The flight had been a triumph, and it owed little to chance.

Meticulous preparation

The *Spirit of St Louis* was an adapted Ryan M-2 mail plane, made to Lindbergh's own specifications. It was constructed of wood and metal tubes covered in fabric; the engine was the most up-to-date version available of the 220hp Wright Whirlwind. Fuel and oil tanks filled the space in front of the pilot – the only windows on the plane opened to the side and Lindbergh had no forward vision except through a periscope. His past experience as a solo mail pilot had prepared him to fly long distances using instruments for navigation; shortly before his Atlantic attempt, he had made a 2,500-mile (4,000km) non-stop flight across the USA.

Lindbergh had taken infinite pains to reduce the payload of his plane, opting to do without a radio, parachute or night-flying equipment. He took only five sandwiches, two bars of chocolate and a bottle of water to sustain him during the flight.

A perfect flight

Delayed at first by bad weather, Lindbergh eventually took off from New York's Roosevelt Field at 7.52 am local time on May 20. He deviated slightly from his planned route to avoid storms over Nova Scotia, passed over Newfoundland, and then took the shortest possible route across the Atlantic Ocean. Throughout the voyage he steered a steady course, overcoming fatigue and all the challenges the plane presented to him. By the time he reached Cherbourg on the French coast, at 8.20pm GMT the following evening, he knew he had pulled it off. When he finally touched down on French soil after a flight of 3,600 miles (5,800km), his fuel tanks still held 322 of the 1,440 litres of fuel that had been loaded before take off.

A hero's reception

Lindbergh was by no means the first man to fly across the North Atlantic. In May 1919 a team led by Albert Cushing Read had made the journey by flying boat in six stages, the longest from Canada to the islands of the Azores. In June that same year John Alcock and Arthur Whitten Brown achieved the first non-stop flight, from Newfoundland to Ireland. Yet even so, Lindbergh's flight seemed to mark a watershed in aviation history. His singular courage caught the world's attention, even though others were striving to do the same. In the three weeks after his landing three other American pilots completed the journey, flying planes similar to the *Spirit of St Louis*. But flying was still a highly dangerous obsession: just 13 days before Lindbergh took off two French aviators, Nungesser and Coli, had been lost with their biplane attempting the same feat in reverse, from Paris to New York.

Lindbergh himself returned to a hero's welcome in New York, parading down Broadway under a snowfall of ticker-tape. Although he could hardly have imagined it at the time, his flight marked the arrival of the American aviation industry on the world stage, which had earlier been dominated by British and French manufacturers. The immense publicity surrounding his achievement attracted US businessmen to invest heavily in the sector, permitting American manufacturers to develop new technologies that would soon outpace their European competitors.

Making aviation history
A portrait of Lindbergh taken at St Louis airfield. He flew from there to New York to launch his intrepid flight.

Hero's welcome
Lindbergh was given a ticker-tape reception on his return to New York (right). He subsequently travelled to Washington, where he was received by US President Calvin Coolidge.

KNIGHTS OF THE AIR

Alcock and Brown were both knighted by King George V after their pioneer non-stop transatlantic flight in a converted Vickers Vimy bomber.

Up and way

The Spirit of St Louis *takes off from San Diego airport in California on 10 May, 1927 (above right). The plane had been made in San Diego, but Lindbergh chose to name it for the city where he and his financial backers lived at the time.*

A CONTROVERSIAL PERSONALITY

After his flight across the Atlantic, Lindbergh (1902–1974) married the writer Anne Morrow, a wealthy heiress. Rich, adored and the father of six children, he nonetheless continued to hanker after the solitude of solo flights. The couple experienced tragedy when their eldest son was kidnapped in 1932 and found dead soon afterwards. Subsequently the Lindbergh family moved first to Kent in England, then to a Breton island. At the start of the Second World War he championed American neutrality and became a spokesman for isolationism, only changing his views in 1941 after the Japanese bombing of Pearl Harbor. Shocked by the horrors of the war, he became in later life a supporter of environmental causes.

The silent screen gets a voice

With the US release of *The Jazz Singer* in 1927, the era of talking pictures got under way. Popular enthusiasm for the 'talkies' revolutionised the film industry.

The first talkie
May McAvoy with Al Jolson wearing blackface make-up in a scene from The Jazz Singer *(below). There was little actual dialogue in the film, but Jolson's songs took the world by storm.*

Al Jolson was at the height of his popularity as a star of vaudeville when, in 1927, he appeared on the big screen in blackface make-up. After singing an introductory number, he quieted the on-screen audience with the words, 'Wait a minute, wait a minute, you ain't heard nothin' yet'. They were the first spoken words to be heard in a full-length feature film. Directed by Alan Crosland for Warner Brothers, *The Jazz Singer* introduced the era of talking cinema. The movie still employed intertitles, just as silent films had done, but the brief interludes of synchronised singing and dialogue marked a revolution in cinema history.

Pioneering experiments

Films had rarely been truly silent before that time; most had been accompanied by some form of live music. Sound recording itself predated the coming of film by a couple of decades. As early as 1895 Thomas Edison had the idea of combining his phonograph with another of his inventions – the kinetoscope, a big box that allowed a spectator to watch moving pictures through a viewing window. But film producers and distributors showed little interest when the popular enthusiasm for silent films showed no signs of waning.

Things began to change in 1924, when Warner Brothers adopted the Vitaphone system, which synchronised recorded discs with projected film. In 1926 they promoted *Don Juan*, which had recorded sound but no spoken dialogue. *The Jazz Singer* followed a year later and proved a triumphal success.

No turning back

Paradoxically, the system used for *The Jazz Singer* was rapidly outdated. It involved recording the images and the soundtrack on separate discs, and synchronising the two proved complex. The real breakthrough came in 1930 when sound was recorded directly onto the film itself, transformed into light waves inscribed in a narrow band that ran alongside the frames. The projection speed was raised to 24 frames a second from the 16 used for silent movies.

The coming of talkies presented problems for the film industry, which had to adapt quickly. Talking pictures were more expensive to make and did not have the universality of silent films. Before the introduction of dubbing in the early 1930s, films for export

ACTORS IN THE FIRING LINE

The coming of the talkies caused problems for actors, who in silent cinema had relied on comportment and gestures to carry the action and express emotion. Many established stars dreaded having to speak in front of the cameras: a nasal voice, a grating accent or poor diction could mean the end of a career. John Gilbert, an icon of the silent screen adored for his macho persona, had a high-pitched voice at odds with his screen image, and he quickly faded from view. In contrast a new star was born in Clarence Brown's 1929 film *Anna Christie*, when an actress imported from Europe addressed a barman in a throaty, sensual voice, demanding: 'Gimme a whiskey, ginger ale on the side. And don't be stingy, baby.' Greta Garbo had spoken her first lines, and a new star was born (below).

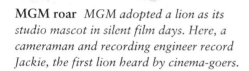

MGM roar *MGM adopted a lion as its studio mascot in silent film days. Here, a cameraman and recording engineer record Jackie, the first lion heard by cinema-goers.*

had to be recorded silently then post-synchronised in different languages in the studio, or were shot in multiple versions.

Some artists, notably Charlie Chaplin, deplored the passing of silent film, with its ambiguities of meaning and the freedom to shoot in the open air. Yet the new technology opened up the potential offered by dialogue and sound effects, as well as whole new genres such as musical comedy, documentaries and psychological thrillers. Alfred Hitchcock's 1929 *Blackmail* was Britain's first feature to incorporate synchronised sound. The public's verdict was unequivocal. From 1930 on, any cinema worthy of the name could not present its audience with anything other than talkies.

Sound track
The width of 35mm film images was reduced to allow room at the side for the audio signal (above).

LIVE RECORDING OR DUBBING

Dubbing made its appearance soon after the advent of the talkies, quickly followed by complaints about the use of inappropriate voices to replace those of the original actors. The critics sometimes forgot that the dialogue heard on screen in most films then as now was not that spoken when the shooting took place; to avoid unwanted interruptions, it was usually re-recorded by the actors afterwards. The popularity of live recording in 1930s films was closely linked to the practice of filming on studio sets rather than on location.

A star is born
Greta Garbo on the set of Anna Christie *(right). She shot to fame for her sultry voice as much as her looks.*

Measuring time by oscillation

Like mechanical clocks before them, quartz clocks were the end product of a long line of separate innovations. The chain stretched from the Curie brothers to Walter Cady, from sonar to radio, and culminated in 1927 in the Bell Laboratories where Warren Marrison and J W Horton produced the first reliable prototype.

The origins of the quartz clock date back to 1890 when the Curie brothers, Pierre and Jacques, discovered that certain so-called 'piezoelectric' materials produce electricity when subjected to pressure or mechanical stress. It soon became apparent that the reverse was also true: that an electric current could cause these materials to vibrate at a rate of thousands of times a second.

In the First World War the American physicist Walter Cady used piezoelectric quartz crystals in his work on sonar underwater detection. From this he learned that the frequency at which the quartz vibrated depended exclusively on its shape; he also found that the rate of vibration was extraordinarily stable and precise. In 1921 he constructed the first quartz resonator to stabilise the frequencies of radio transmissions.

Technological breakthroughs

Cady's resonator device aroused a great deal of interest in scientific circles. Soon numerous research teams were seeking to take advantage of its unique qualities, which made it ten times more accurate than the finest astronomical clocks available at the time. In 1927 Warren Marrison and J W Horton, a pair of telecommunications engineers at Bell Laboratories in New York, became the first to capitalise on Cady's work for time-keeping purposes. They produced a quartz resonator that vibrated 50,000 times per second, with an electronic circuit that counted the vibrations. Each time the requisite number was reached, an impulse was sent to an electric motor that move the clock's second hand forward. The clock was so accurate it lost just one second in three years. There were still problems to overcome: the clock was susceptible to changes in temperature – a swing of 20°C (36°F) could cause it to lose two seconds a day – and it was the size of a large fridge-freezer.

In 1932 the extraordinary precision of the quartz clocks enabled scientists to measure

Quartz-crystal electronic oscillator
As a result of the piezoelectric effect, electrodes set on a quartz strip cause the crystal prongs to vibrate like a tuning fork. These pink quartz crystals (below) take their colour from manganese or titanium oxides.

SYNTHETIC QUARTZ

Every year 2 billion quartz crystals are produced synthetically by recrystallising natural quartz to increase its chemical purity. Correctly, the end product is referred to as silicon dioxide. Shaped like tuning-forks and assembled in a vacuum to increase their precision, watch crystals vibrate 32,768 times a second.

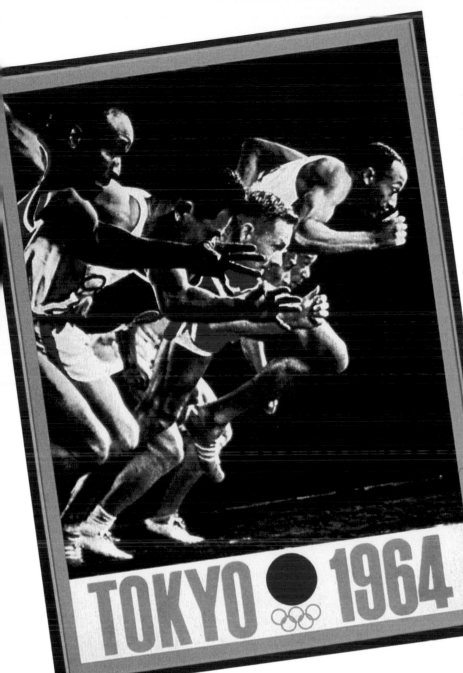

TOKYO ● 1964

Quartz devices quickly came to dominate the world of sports time-keeping, setting off a technological race to see who could develop the first watch using the new technology.

Seiko succeeded in putting the Astron on the market in late 1969, but the first models had a price-tag that matched that of a small car. The firm's Swiss competitors soon followed suit, but even though prices fell, quartz watches continued to be luxury products that were made by hand. The mechanisation of quartz production and the introduction of liquid-crystal displays – by the Japanese firm Suncrux in 1975 – finally made products available at affordable prices that were accurate to a thousandth of a second a day. Such precision has become indispensable to many of the electronic gadgets that have since become part of everyday life: mobile phones, processors, CD players and wi-fi, for example, all contain tiny quartz components.

TECHNOLOGICAL RIVALS

From the 1950s on, even before the advent of quartz mechanisms, the Swiss watch industry faced a serious challenge from Japanese competition. In response, a number of top firms formed a consortium in 1965, hoping to make up for lost time by launching a joint research project into the new technology. A prototype, the Beta 1, was assembled and tested two years later. The first commercial Swiss quartz watches reached the market in April 1970, four months after the Seiko Astron.

Historic prototype
Seiko unveiled the Astron watch (above) in 1969. A hundred examples made of gold were sold in the first week.

miniscule variations in the Earth's rotation. They remained the criterion of accuracy until 1956, when the first commercial atomic clocks challenged their pre-eminence with a time lag of just one second in a million years. The new devices still used quartz resonators, but owed their improved efficiency to elements such as caesium measuring the emission of electrons.

A thousandth of a second a day

For years quartz clocks were cumbersome objects, confined to scientific and technological uses. That changed at the 1964 Olympics, when the Japanese firm Seiko unveiled the first transportable quartz stop-watches, which were accurate to one hundredth of a second. Thanks to thermal controls, amplification and the use of transistors to modulate the current, the size had been reduced to just 16 x 20cm (6 x 8 in).

Time-keepers for champions
Quartz-crystal chronometers were introduced at the Tokyo Olympic Games of 1964, when Seiko's QC-951 model was used to time some of the events. Quartz technology had come a long way since Marrison and Horton's first clock, but the Japanese firm was the first to resolve the problem of miniaturising the components.

Adhesive tape 1928

In 1925 Richard Drew, a laboratory assistant employed by the American 3M manufacturing company, was working on a commission for an automobile manufacturer. At the time fashion dictated that cars should have two-tone finishes, and to make the job of the painters easier he came up with the concept of masking tape, produced by vaporising glue on a cellulose strip. The idea was good, but at first the tape often failed to hold for want of sufficient adhesive. Blaming the fault on the firm's stinginess, purchasers jokingly christened the product 'Scotch tape', drawing on the pejorative stereotype of the Scots and their supposed financial canniness.

Three years later Drew put the Scotch tape we know today on the market. It quickly found buyers among people seeking to repair tears in paper, fix sheets together or attach documents to a support. In 1932 John A Borden, a colleague of the inventor, developed the familiar roll format. Since that time adhesive tapes have continued to flourish, and today there are more than 900 different brands.

ELASTOPLAST

Elastoplast was introduced by the English pharmaceutical firm Smith & Nephew in 1928. Taking advantage of the development of water-resistant adhesive tissue, they devised a new form of wound dressing consisting of a piece of gauze coated in an antiseptic solution attached to a strip plaster. Since then various different types of sticking plaster have been created to suit every kind of minor cut or graze.

Mass-produced yogurt 1929

In 1919 Isaac Carasso, a Spanish doctor and entrepreneur, started producing home-made yogurt using fermentation agents supplied by the Institut Pasteur in France. He set up a firm in Barcelona to produce the yogurt, which he initially sold in pharmacies. He called the company Danone, from 'Danon', the family nickname for his son Daniel.

Ten years later Daniel himself opened the company's first French factory in the northwest suburbs of Paris, introducing industrial techniques to supply a mass market. In 1941 the family, who were Jewish, moved to New York following the German occupation of France. They returned a decade later and in 1983 Daniel created an international centre for research into gut flora and the beneficial micro-organisms known as probiotics.

Yogurt production line
Danone introduced the first fruit yogurt in 1937.

The yo-yo 1929

Composed of twin discs joined by an axle with a piece of string wrapped around it, the yo-yo could well be the world's second oldest toy after the spinning top. Known in China around 1,000 BC and in ancient Greece around 500 BC, it arrived in Britain at the end of the 18th century, when it was known as the bandalore. In 1928 a Filipino-born US citizen named Pedro Flores started to mass-produced the toys as yo-yos, which was their name in Tagalog, the local language in the Philippines. The following year, the American entrepreneur Donald Duncan bought up the rights, trademarking the term 'yo-yo'. Duncan also introduced a nationwide series of contests for users. There were significant yo-yo revivals in the 1960s and then again in 1999, the latter partly inspired by the Japanese Bandai company.

In-flight refuelling 1929

Strictly speaking, the first in-flight refuelling took place in November 1921, when two stuntmen managed to transfer a can of gasoline from one biplane to another. It was 1929, though, before two aircraft effectively succeeded in replicating that feat, when the fuel tank of a triple-engined Fokker engaged in an endurance trial was filled via a hose from a Douglas C1 of the US Army Air Corps. Aircraft today still employ essentially the same technique. The tanker plane either has a retractable rigid perch that can be deployed by an operator to dock with a nozzle attached to the receiver's fuselage, or else with one or more pipelines with funnel attachments that fit into a retractable boom attached to the plane to be refuelled. In-flight refuelling became an essential tool in wartime, increasing the range of fighters and bombers and keeping planes airborne over enemy territory.

Tanker and receiver
An aerial tanker refuels a three-engined Fokker above Culver City, California, in 1929. The Fokker's pilots, Loren Mendell and Pete Reinhart, created a new world endurance record by remaining airborne for 246 hours and 43 minutes.

THE SUPPLY PLANES

Most supply planes (known as tankers) were former civil airliners capable of carrying substantial fuel loads over long distances. They all had the same mission: to refuel other aircraft (the receivers) in flight.

EDWIN HUBBLE – 1889 TO 1953
Champion of astronomy

The astronomer Allan Sandage said of Edwin Hubble that he 'solved four of the central problems in cosmology, any one of which would have guaranteed him a position of the first rank in history'. His achievements made him one of the founders of modern astronomy.

Revelations from deep space
A photograph taken by the Hubble Space Telescope shows a cloud of dust and gas in the Carina Nebula, the result of an explosion of a star sometime in the 19th century. New stars will be born from the chaotic energy generated by the blast. Locating the telescope in space allows it to capture images of such astonishing clarity.

By all measures, Edwin Hubble is recognised as a giant in the field of astronomy with major discoveries to his credit. Between 1923 and 1925 he proved the existence of other galaxies besides our own in the universe, putting an end to a controversy that had long divided astronomers. In 1926 he drew up the first morphological classification of the galaxies, establishing the three principal categories: spiral, elliptical and irregular. Over the next ten years he studied the distribution of galaxies across the heavens with the aim of measuring the spatial curvature predicted by Einstein's theory of relativity – work that would pave the way for succeeding generations to discover the large-scale structure of the universe. Above all he proved in 1929 that the galaxies are pulling away from one another, thereby demonstrating that the universe is expanding, a discovery that finally confirmed the Big Bang theory that has since come to dominate cosmological thinking.

Late starter

Hubble's name is associated with at least 11 separate laws, principles or astronomical objects, most famously the Hubble Space Telescope which went into service in 1990. It is an impressive record for a middle-class native of Marshfield, Missouri, who hesitated

Eye on the stars
Edwin Hubble (left) with the gigantic Hale Telescope in 1949, the year that the telescope came into operation at the Palomar Observatory in California. Hubble was the first astronomer allowed to use it.

between science, languages, law and sport, for which he showed considerable ability, before starting his career in astronomy at the age of 28, when he gained a PhD in the subject. While working on his doctoral thesis – 'Photographic Investigations of Faint Nebulae' – he developed the techniques of deep-space observation that he would use for the rest of his life.

THE EXPANDING UNIVERSE

In 1929 Hubble, aided by fellow American Milton Humason, published his law relating to the rate of expansion of cosmic space based on observations of red shift. The light given off by a source moving away from the Earth appears redder the faster it is travelling. By observing the luminosity of Cepheid stars in 46 different galaxies, Hubble reached the conclusion that the speed at which they were moving followed a formula: a galaxy 1 light-year away moved at a rate of 153km (85 miles) a second, at 2 light-years away 306km (170 miles) a second, and so on. Today, thanks to improved measurement of stellar distances, the accepted speed has been reduced to about 70km (43 miles) a second per light year, but Hubble's principle still stands.

Star-gazer supreme

Hubble's expertise won him a position in 1919 at Mount Wilson Observatory in California, home to the first giant telescope of the 20th century, equipped with a mirror 2.5m in diameter. There this pragmatic genius set the standard for astronomical observation in the modern era. He used the instrument to scrutinise and photograph the heavens while applying the most up-to-date calculation methods, including measuring red shift in nebulae and galaxies. Having found his spiritual home at Mount Wilson and his intellectual home in the study of nebulae, he remained faithful to both for the rest of his life. Hubble died in 1953, without the ultimate accolade of a Nobel prize as none were available in his field. In his honour, the Nobel prize rules were changed soon after to make astronomers eligible for the physics award.

A meeting of galaxies
Stephan's Quintet in the constellation of Pegasus – seen here in images from the Hubble Space Telescope – is a grouping of five galaxies, two of them elliptical and three barred spirals. All five may eventually merge; the two at the centre of the picture are already coming into contact with one another.

The Hubble sequence
A diagram published in 1936 illustrates Hubble's conviction that galaxies evolve over the course of astronomical time – as shown from left to right above – moving from spheroid (E0) to elliptical (E3, E7), then to a lens shape (S0) before becoming either normal (Sa, Sb, Sc) or barred spiral (SBa, SBb, SBc). With some qualifications, the sequence is still used by astronomers today.

GALAXIES BEYOND GALAXIES

Does the Universe extend beyond the Milky Way? This question had long divided astronomers – until Hubble provided a definitive answer by showing that some nebulae were actually distant galaxies lying far outside the bounds of our galaxy. He did so with the help of a scientific law, established in the 1910s by American astronomer Henrietta Leavitt, for determining the distance of Cepheids – a class of regularly pulsating variable stars – by measuring the amount of light they gave off. By applying this formula to a variety of nebulae, including NGC 6822, M33, M32 and M31, Hubble was able to show that their distance from Earth far surpassed the Milky Way's diameter.

The miraculous mould

Alexander Fleming published his first article on penicillin in 1929, but it was only with the coming of the Second World War that the miracle antibiotic was produced in industrial quantities, saving the lives of many thousands of people.

'I have been accused of having invented penicillin. No man could invent penicillin, for it has been produced from time immemorial by a certain mould. No, I did not invent the substance penicillin, but I drew people's attention to it, and gave it its name.' These are the words of Alexander Fleming, speaking at a ceremony held in Brussels in 1945 to honour his discovery. A few weeks later, the Scottish bacteriologist was awarded the Nobel prize for medicine jointly with Howard Florey and Ernest Boris Chain, two colleagues in the pathology department at Oxford University, who had found a way to harness the antibiotic properties of the fungus for practical medical purposes.

Revelation through contamination

The saga of one of the greatest discoveries of the 20th century began in 1928, when Alexander Fleming was working as a bacteriologist in London at St Mary's Hospital, Paddington. He was studying the antibacterial properties of lysozyme, an enzyme he had traced in tears and nasal secretions. In early September 1928 Fleming returned from holiday to an unwelcome surprise: a bacterial culture that he had left uncovered on a bench in his laboratory had been contaminated by a blue-green mould.

Taking a closer look at the Petri dish containing the culture, Fleming realised that the colonies of staphylococci which he had been carefully nurturing had been destroyed in the immediate vicinity of the mould. With the aid of a microscope, he was able to identify the mould as *Penicillium notatum*, a fungus that a mycologist colleague happened to be studying at the time. Penicillin, as Fleming named the mysterious substance, had just been discovered by accident.

Medical memento
A souvenir display case containing a sample of Alexander Fleming's penicillin mould (below).

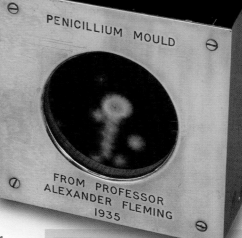

Cultivating penicillin
The Penicillium chrysogenum *strain of penicillin (above) is relatively easy to cultivate in quantity. Mutant strains have been developed that can deliver in excess of 60,000 units per millilitre.*

PENICILLIN'S PRECURSORS

Honey, herbs, moulds ... Ever since antiquity, doctors and healers have sought out natural substances to combat infections on a kill-or-cure basis. Some 2,500 years ago Chinese physicians were already using soya unguents covered in mould to treat skin infections. In the 1st century AD the Greek doctor Dioscorides recommended the use of yeasts for purulent wounds. The concept of antibiotic therapy – in the sense of using certain classes of micro-organisms to destroy others – was a product of the 19th century, born in the wake of Louis Pasteur's researches. In about 1885, for example, an Italian named Arnoldo Cantini suggested spraying harmless *Bacterium termo* in the mouth of patients suffering from tuberculosis to combat the agents causing the disease.

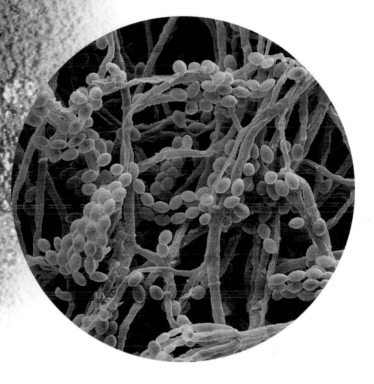

Beneficial growth
In its natural state (below), penicillin mould is a tiny fungus that grows on mouldy fruit and some cheeses, notably Roquefort.

In fact, Fleming was not the first person to become aware of the special properties of penicillin. In 1876 the physicist John Tyndall had discovered that strains of *Penicillium* inhibited the growth of a variety of germs. At the end of the 19th century a French doctor, Ernest Duchesne, had noted that the mould killed certain bacteria.

Proving penicillin worked

Having successfully tested penicillin solutions on streptococci, meningococci and bacilli, Fleming published his findings in a medical journal in 1929. There was very little immediate response and even he did not put much faith in the future of penicillin as an antibiotic, the more so because it had proved difficult to cultivate and isolate. Fleming gave up work on the substance altogether in 1932, by which time penicillin was merely being recommended for use as a chemical reagent and as a surface

THE SULPHONAMIDES

The first recorded occasion on which an antibiotic healed an infectious disease was in 1932, when an infant covered with wounds and abscesses and with a temperature of 40°C (104°F) was saved from death by the first sulphonamide drug, Prontosil. Made by synthesising the crystalline dye chrysoidine with a sulphur-based derivative, the compound was discovered by a German doctor, Gerhard Domagk. It proved effective against streptococcal and staphylococcal infections, both endemic at the time. Other sulphonamides that countered different bacteria soon became available. The star of these was Bactrim, launched in the 1970s; this powerful antibiotic was found to help suppress ailments such as toxoplasmosis and pneumocystosis that often accompany HIV/AIDS.

Pioneer of sulphonamides
Shown here in a portrait by Otto Dix, Doctor Gerhard Domagk won the 1939 Nobel prize for medicine, but was prevented from accepting it by Germany's Nazi authorities. He was finally able to receive the award in 1947.

Producing penicillin
A photograph of 1943 shows row upon row of flasks containing Penicillium *cultures; each would yield penicillin after three weeks' cultivation.*

antiseptic. And this is how it remained until the end of the decade, when Howard Florey and Ernest Boris Chain, working together at Oxford University, managed to find a way of mass-producing the active ingredient of the penicillin mould. In experiments conducted in 1940 they showed that mice infected with lethal doses of streptococci could be saved by counteractive doses of penicillin. The first trials on human patients were carried out in 1941 and were spectacularly successful.

From trials to saving lives

Penicillin had been proven to work, but the supply of the drug was so limited that the urine of patients treated was filtered to recover the antibiotic for re-use. At the height of the Second World War, Florey travelled to the USA to try to persuade manufacturers to produce the drug on a grand scale. Several laboratories flung themselves into production using a new cultivation method. Penicillin manufacture became part of the war effort, alongside that of steel, and many Allied soldiers would owe their lives to the new wonder drug.

When peace was restored penicillin became available from pharmacies and was soon prescribed for a whole range of infections. Although the results were sometimes amazing, the antibiotic proved ineffective against some bacteria and it could only be administered by injection. Pharmacists gradually developed a derivative that could be taken orally and that proved efficacious against other germs.

Meanwhile, other families of antibiotics were emerging, paromomycin, the macrolides

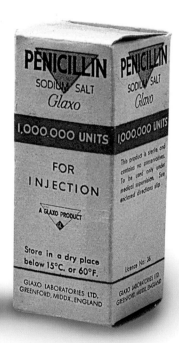

Mass-market product
A flask of injectable liquid penicillin (right) dating from about 1950. The manufacturer was Glaxo, a firm that dominated the market in the drug in the years after the Second World War.

BACTERICIDES AND BACTERIOSTATICS

Antibiotics divide into two principal groups: bactericides, which destroy germs, and bacteriostatics, which prevent germs from multiplying. Scientists have revealed the mechanisms by which they operate. Some attack the structure of bacterial cell walls, while others inhibit the germs from replicating their DNA or synthesising proteins.

and tetracyclines among them. The hundred or so agents available today have contributed greatly to the reduction in infant mortality and increase in life expectancy experienced in the latter part of the 20th century. Yet the intensive use of penicillin and other antibiotics has triggered concerns about the growing resistance of germs. Attempts to counteract this, mostly by limiting the prescription of antibiotic drugs, have become a priority for medical authorities hopeful of preserving the efficacy of these precious weapons against bacteria that threaten human health.

A NINE-YEAR JOURNEY

The first space probe to visit Pluto is due to arrive in its orbit in July 2015, more than three-quarters of a century after the discovery of the dwarf planet on the outer reaches of the Solar System. The New Horizons robotic craft, launched by NASA in 2006, aims to gather and send back information on the temperature, composition, morphology and atmosphere of Pluto and of its largest moon, Charon. After almost a decade of travelling, the flyby will take less than 24 hours.

Pluto and Charon
An infrared image of Pluto and its chief moon, Charon, taken from the Gemini North telescope in Hawaii in June 1999.

Pluto 1930

Pluto's discovery followed on from that of Uranus in 1781 and of Neptune in 1846. Even taking into account the gravitational pull of Neptune, the orbit of Uranus did not conform to what could have been expected from Newton's laws. Astronomers therefore came to the conclusion that there must be a ninth planet whose presence would account for the anomalies.

The mystery object was finally tracked down in 1930 by a self-taught American astronomer named Clyde Tombaugh. He spent more than a year at the Lowell Observatory in Arizona poring over photographic plates taken at intervals of several days. He used a device called a blink comparator, which allows rapid switching between images of the same area of sky, thereby making it easier to spot any changes in the position of objects. His tenacity enabled him to flush out what he was looking for from the 300,000 or so stars whose brightness tended to overshadow the other heavenly bodies on the plates.

Orbiting on average 5.9 billion kilometres from the Sun, the newcomer was barely visible from Earth. It was given the name of Pluto. In 1978 the discovery of Charon, Pluto's largest moon, which periodically passes across Pluto's face, enabled astronomers to calculate the diameter of what was then regarded as the Solar System's ninth planet. The figure came out at just 2,340km (1,450 miles).

Pluto consists of bare rock and methane ice and has just 0.3 per cent of the Earth's mass. The 1990s saw the discovery of what amounted to a ring around Pluto's neighbour Neptune, made up of a thousand or more bodies rivalling Pluto in size. These so-called 'trans-Neptunian objects' subsequently led the International Astronomical Union, in 2006, to downgrade the classification of Pluto from planet to dwarf planet.

Too small for a planet
Orbiting the Sun at the far reaches of the Solar System, Pluto – just visible at top left in the image above – is now classed as a dwarf planet, which means that it is large enough to have been rounded by its own gravity.

Using infrared cameras to detect hotspots

Work done by a Danish professor in Copenhagen opened the door to thermal imaging. The technique has since proved to have three main applications: in medicine, for detecting tumours and areas of infection; in industry, for checking out heat loss or overheating; and in construction, for evaluating the effectiveness of thermal insulation.

Rendering temperature visible
Infrared photography registers temperature variations as different colours, ranging from the lowest (in black) to the highest (white). This image of a patient suffering a severe migraine (above) shows the hottest areas at the base of the neck, around the eyes and in the cranial region.

At the start of the 1930s, a professor by the name of Haxthausen at the University Hospital in Copenhagen decided to use infrared photography to examine human skin. Infrared had been discovered in 1800 by the astronomer William Herschel, but more than a century later this luminous form of radiation, invisible to the naked eye, had only begun to be exploited photographically. Haxthausen used the technique to reveal subcutaneous veins that could not otherwise have been seen in a clinical examination.

Despite these promising beginnings, over the next few years little use was made of infrared photography in medicine for want of suitable equipment. A military use emerged in the Second World War to detect the presence of hostile troops wearing camouflage, for infrared images can differentiate easily between warm-blooded humans, however well disguised, and surrounding vegetation. The first colour images were produced at this time, in 1942, employing Kodak film.

Diagnosing breast tumours
Medical thermography finally got off the ground in 1957, when a Canadian doctor named Ray Lawson showed that the temperature of cancerous skin is higher than normal as a result of inflammation and the

---

INFRARED PHOTOGRAPHY

The first infrared photograph was taken in 1880 by a Derby-born chemist named William Abney, who used a collodion-based emulsion to capture an image of a hot teapot. But what Abney did not do was pass on how he had done it. Subsequent attempts at thermal imaging proved relatively fruitless until the discovery, in 1903, of dicyanin, a colouring agent sensitive to infrared light. From 1910 on, the American physicist Robert W Wood used an allied substance, cryptocyanine, to take and publish black-and-white thermal photographs of landscapes.

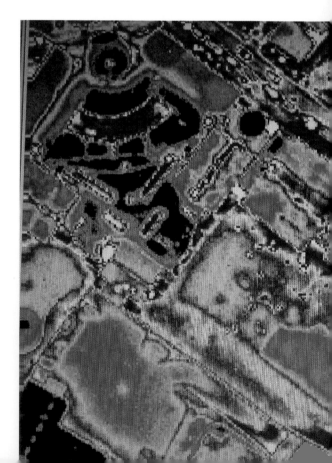

development of new blood vessels around tumours. In fact, thermal cameras do not so much photograph variations in body temperature directly as the infrared rays given off by the patient. Like any other object, the human body emits electromagnetic waves whose length reflects temperature. Infrared cameras can identify hotspots.

In 1980 a French medical team under the direction of Yves Leroy developed the more sophisticated and sensitive technique of microwave imaging, which can detect temperature variations down to a tenth of one degree Celsius without exposing patients

Heat-seeking machine
Installed on the US-Mexican border, infrared telescopes like this one (above) help the security patrols to spot the activity of illegal immigrants.

to radiation. Thermal imaging techniques received official recognition for breast cancer screening in the USA in 1982. Over the course of the 1980s they became popular in dermatology, rheumatology and neurology, and were also used to examine blood vessels.

Fresh uses for heat detectors

In the years that followed infrared photography gradually fell out of favour in medicine, overtaken by more effective radiological techniques. Yet it continued to be used in one particular area: for airport detection of infected travellers returning from quarantine regions struck by epidemics.

In contrast, thermography played an increasing role in other fields. In industry thermographic cameras have become popular detection devices, being used to spot heat leaks or overheating. They are also widely used in construction. Some European municipal authorities have turned to thermographic aerial photography as a way of identifying heat loss from roofs or poorly insulated areas in blocks of flats and public buildings. Individual householders can also commission thermal checks on their homes to combat energy loss, thereby doing their bit in the struggle against global warming.

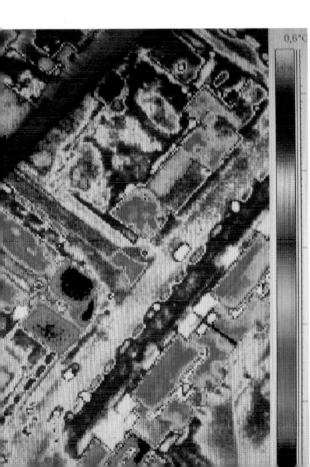

Thermal imaging
An aerial photograph of a school reveals the thermal efficiency of the buildings. The blue areas indicate buildings that suffer the least heat loss, while white spots highlight places where insulation is inadequate.

The electric organ 1931

A precursor to the electric organ, known as the telharmonium or the dynamophone, was unveiled in 1897. Invented by an American, Thadeus Cahill, it had one major disadvantage: it weighed 7 tonnes. In 1927 two French engineers, Eloy Coupleux and Armand Givelet, developed a 'wave organ', first heard in public in 1929. The principle was simple: a series of oscillators produced the notes, or frequencies, while filters varied the timbres. A limited number of wave organs were produced from 1931 on.

Three years later a clockmaker named Laurens Hammond took out a US patent on a revolutionary new design. He employed high-precision tonewheels, rotating in front of electromagnetic pick-ups, to emit frequencies that jointly produced the sound. Improved over the years, the Hammond organ became a staple of churches and cinemas.

Electric organs paved the way for synthesisers, which were first introduced by another American, Robert Morgan, in 1964. They in their turn became standard features of pop music and film soundtracks.

Playing the organ
An early Hammond, photographed in the 1930s, featuring two keyboards with 61 keys and a 25-note pedalboard.

The electric guitar 1931

The case for amplifying guitars electronically was clear enough, for guitar-players had difficulty making themselves heard within big bands. In 1931 three Californians addressed the problem. Paul Barth and George Beauchamp devised a means of transforming the vibrations of each string into electric currents that could then be amplified. Guitar-maker Adolphe Rickenbacker set about commercially exploiting the invention, and other guitar manufacturers soon followed suit.

In the 1940s the American jazz guitarist Les Paul contributed many improvements, notably the new solid-body guitar. By doing away with the sound box, this helped to eliminate unwanted feedback heard in the form of high-pitched screeches and whistles.

In a league of his own
Jimi Hendrix photographed on stage at the Royal Albert Hall in London on 24 February, 1969. Hendrix, who was left-handed, is playing a right-handed instrument with the strings reversed.

THE ELECTRIC REVOLUTION

Electric guitars became the chosen instrument of rock'n'roll guitarists. Chuck Berry with his Gibson 335 in the 1950s and a decade later Jimi Hendrix, who favoured Fender Stratocasters, were two of the principal innovators. In jazz, electronic amplification turned guitarists into star soloists, as Charlie Christian with his Gibson ES-150 and Wes Montgomery, mostly using the L5-CES model, were memorably to demonstrate.

Tampons 1931

Sanitary napkins, introduced by Kimberly, Clark & Co in 1920, were a major step forward in their day, offering an alternative to home-made menstrual towels. The next advance came when a doctor named Earle Haas had the idea of applying the principle of the surgical tampon used to staunch a haemorrhage to menstruation. Haas patented his invention in 1931, incorporating such ingenious details as an aseptic applicator tube that aided insertion and a cord attached to the tampon to make it easier to remove. Two years later, having failed to find a backer, he sold the rights to his invention to a businesswoman called Gertrude Tenderich. She launched the Tampax (short for 'tampon packs') brand in 1936.

Long-playing records 1931

In 1931 two Americans named Harrison and Frederick came up with an idea for improving the crackly sound quality of 78-rpm records. They proposed cutting discs from vinyl, a softer material than the brittle shellac that had previously been used, and also playing back with a lighter cartridge head. In 1933 other inventors made further improvements that included narrowing the grooves to little more than half their previous size and using a cartridge with a stylus able to 'read' sounds from the sides of the groove. The invention may have been good but the timing was bad, for the economic crisis of the 1930s discouraged record companies from jettisoning their old equipment. RCA Victor gave up producing the discs in 1933.

Fifteen years passed before Peter Goldmark of Columbia Records revived the idea. Further narrowing of the groove gave the new records a playing time of more than 20 minutes, compared with the four-and-a-half minute maximum of 78s. In 1948 Goldmark patented the discs, which played at 33⅓ revolutions per minute and soon became known as LPs (short for 'long-playing'). Cheaper than 78s for the amount of music they contained and offering a better sound quality, LPs gained an increasing share of the market, only to be dethroned after 1979 by compact discs, whose digital recordings were read by laser beam.

In the groove
With analogue LPs the stylus – shown below in a much-magnified image – transformed vibrations that varied with the depth of the groove on the record into electronic signals amplified by loudspeakers.

DJ at the turntable
Disc jockeys gained cult followings in the 1980s, bringing fresh life to the vinyl market.

STEREO RECORDS

The first stereo 78s, employing a system devised by the physicist Alan Dower Blumlein, were unsuccessfully launched by EMI in 1933. Twenty-five years later Audio Fidelity in the USA and Pye and Decca in Britain introduced the first stereo LPs. To record them, two microphones, each linked to an amplifier, produced separate soundtracks, captured simultaneously and pressed onto different sides of the grooves. For playback the stylus had to move from side to side as well as up and down.

Casting new light on the incredibly small

As early as 1873, the German optometrist Ernst Abbe had demonstrated that the problem of light diffraction prevented miniscule objects from showing up clearly under optical microscopes. In 1931 two German researchers replaced the ray of light in the microscope with a beam of electrons, opening a new route for exploring the world of the infinitesimally small.

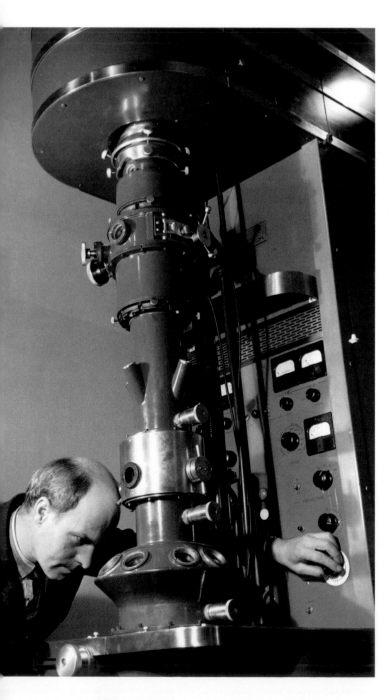

In Berlin in 1929 a student engineer named Ernst Ruska was working on a cathode-ray tube project. There was nothing very new about the task that his professor, Max Knoll, had set for him. It involved increasing the intensity of the electron beam inside the tube, but Ruska went further. He learned that his fellow-German Hans Busch had put forward the idea that a magnetic field could concentrate a particle beam much as an optical lens does light. To achieve the effect the field needed a precise form, which Busch had calculated, but he had not found a way to test his hypothesis.

Ruska had the idea of modifying the electromagnetic coil, habitually employed in cathode-ray tubes, to fit the bill. He covered it in steel, leaving only a narrow aperture on the interior surface, and carried out the necessary measurements so that the reduced magnetic field that emerged from the gap had Busch's suggested shape. In 1929 Ruska patented this 'magnetic lens', and the devices continue to be used to this day.

ABBE'S LAW

In 1873 Ernst Abbe, a German optometrist, demonstrated that diffraction makes it impossible to focus separately on two points whose distance from one another is less than half that of the wavelength involved in viewing them. For optical microscopes using visible light, that equates to a maximum resolution of 0.2 microns (millionths of a metre). Electrons also have a wavelength, but one that is on average 100,000 times shorter than that of light, with the result that the resolution of an electron microscope can go as low as 0.1 billionth of a metre.

Early exemplar
A researcher using an early electron microscope at the National Physical Laboratory at Teddington, south-west London, on the eve of the Second World War (far left).

the objective lens. His first prototype, made in 1931, had a magnification of only 15x, but nonetheless established the principle on which electron microscopy would be based. By November 1933 Ruska had achieved 12,000x magnification, ten times better than the best optical microscopes of the day. In 1939, when the Siemens engineering firm introduced the first commercial model, the order of magnification had increased to 30,000x.

A multitude of uses

New fields of investigation rapidly opened up. In medicine, from 1940 on, it became possible to observe viruses, which had previously been difficult to study as they were too small to be seen under conventional microscopes. In biology, analysis revealed the functions of the constituent parts of cells. Metallurgists explored the world of crystallographic defects, paving the way for the engineering of more solid materials.

The instruments themselves have been constantly improved, leading to the creation of a range of different types, each with specialised applications: diffraction microscopes for crystallography, spectroscopic microscopes for *in situ* chemical analysis of microscopic particles, and so on. The entire electronics industry, including the manufacture of processors and hard discs, could not have developed as it did without the quality control made possible by the 3D images provided by scanning electronic microscopes. The same holds true for nanotechnology.

Meanwhile, the power of magnification has continued to expand. High-resolution transmission microscopy techniques, employing software to correct aberrations, have now taken the degree of enlargement beyond 50 million, enabling scientists to study atomic structure directly.

Scanned images
Close-up images of fat tissue (top) and calcium phosphate crystals (right) taken with a scanning electron microscope. Biological specimens are often given an ultra-thin heavy-metal conductive coating to improve contrast and reduce static.

Smaller and smaller

The notion of a lens soon suggested the possibility of images. Ruska and Knoll duly adapted a long cathode-ray tube to take the first images ever captured with the aid of electrons. Quickly seeing ways to improve the device, they realised that their prototype electron microscope could easily be made to outperform optical microscopes, which Ernst Abbe had shown were limited to roughly 1,000x magnification.

After two years of relatively little progress, Ruska had the idea of linking two magnetic lenses, one to take the part of the eyepiece in an optical microscope and the other to replace

TRANSMITTERS AND SCANNERS

In transmission electron microscopes (TEMs), the electron beam passes through the specimen to be studied, taking with it information that the objective lens then magnifies for observation. In scanning electron microscopes (SEMs), the beam sweeps successively across rectangular areas of the specimen, building up a three-dimensional image of the whole. Scanning transmission electron microscopes (STEMs) combine features of both methods.

Delicate work
A researcher (left) prepares a specimen for examination through a scanning electron microscope. The reconstructed image shows up on a television screen.

THE CENTRAL ASIAN MISSION
By half-track along the Silk Road

Trade caravans had plied the Silk Road between China and the Mediterranean since the 2nd century BC. In the early 1930s, French automobile manufacturer André Citroën determined to follow in their footsteps to promote the reliability of his firm's vehicles.

The only way
The leader of the Silk Road expedition was Georges-Marie Haardt (standing, above). On 25 July, 1931, after crossing the Gilgit Pass, he was forced to have the half-track vehicles taken apart. The road ahead had disappeared, carried away by a massive avalanche. The only way to get the vehicles down the narrow track that remained was to carry them in pieces. They were reassembled two days later on the far side of the slip.

André Citroën, the flamboyant boss of the Citroën motor company, believed passionately in 'development and global expansion by means of the automobile'. He dreamed up the Central Asian Mission as a way to win international attention for Citroën and appointed Georges-Marie Haardt, his close friend and right-hand man, to lead the trek. The project drew on memories of early endurance tests, such as the Peking-to-Paris motor race staged in 1907, and the French press of the day quickly nicknamed it the *Croisière Jaune* ('Yellow Cruise')

A continent clouded in mystery

The projected route linked Beirut to Beijing by way of Syria, Iraq, Persia and Russian Turkestan, before following the south coast of the Caspian Sea to Xinjang then crossing the Gobi Desert. But three months before the start date the USSR withdrew permission for the vehicles to travel through Soviet territory.

Forced to change his plans, Haardt split the team into two separate groups. One, known as Pamir, came under the control of Haart himself and his deputy, Louis Audouin-Dubreuil. It was equipped with seven half-track vehicles powered by 1,600cc, 30hp motors, each of which could be easily dismantled if required. From the Lebanese capital this group was to head for Chinese Turkestan by way of Persia, Afghanistan and northern India. The other group – called China and led by a naval lieutenant, Victor Point, who had obtained permission from Chiang Kai-shek to travel through Chinese territory – was to set out westward from near Beijing. They were equipped with heavier military vehicles with six-cylinder, 2,500cc engines producing 42hp and had specially designed front tyres as well as supple caterpillar tracks at the rear.

These two motorised caravans, each equipped with a wireless transmitter and receiver, planned to meet up in Kashgar, at the foot of the Himalayas in western China, on 14 July, 1931, before making tracks for Beijing's Forbidden City. In all, 43 individuals were involved in what André Citroën insisted was not just an endurance test-cum-publicity stunt but rather 'a French scientific and cultural mission through Asia'.

In the shadow of the Himalayas

The Pamir group left Beirut on 4 April, 1931, and two days later the China contingent set

out from the port of Tianjin, 120km (75 miles) from Beijing. Over the next three months the eastbound column passed through Damascus, Baghdad, Teheran, Kabul and Peshawar, picking up fresh supplies of petrol, oil and provisions at each stop. They travelled on to Srinagar in Kashmir at the foot of the Himalayas, the principle obstacle in their path, where they prepared to wend their way up to the roof of the world. Only two of the vehicles, specially lightened of their loads, undertook the trek to the Gilgit Pass, 4,750m (15,500ft) up on the border between what is now Pakistan and China.

The intrepid travellers encountered problem after problem. Sometimes they had to blast a way through with dynamite. The mule-tracks they were following were often so narrow that vehicles were left hanging over the void – the slightest error could plunge them to destruction. On 45 separate occasions the bridges were too weak to take the weight of the vehicles, which had to be emptied, dismantled, then transported by cable in pieces to the other side, where they were laboriously reassembled. On two occasions, the dismantled vehicles had to be carried for hundreds of miles by porters. Tested to the limit, both half-tracks and humans laboured under the strain but in the end overcame all obstacles.

The Pamir contingent finally reached the far side of the Gilgit Pass on 4 August. There they learned that their colleagues in the China group had been detained in Ürümqi by rebels who had seized control of Xinjiang province. On hearing this disturbing news, Haardt abandoned the half-tracks and set out on

INVENTOR OF THE HALF-TRACK

The success of the Citroën expeditions owed much to a French engineer named Adolphe Kégresse. As a young man he had gone to Russia to take up a position looking after the Tsar's motor fleet. While there Kégresse invented the caterpillar track to improve the traction on snow. Returning to France after the 1917 revolution, he joined Citroën, where he was put in charge of the all-terrain department. There he developed the half-track vehicles, provided with wheels in front and caterpillar tracks behind, that were used for expeditions first to Africa and then Asia.

Hard going
The westward-bound China group creeps through a gorge near Toksu in eastern Xinjiang (above) and makes its way across the Gobi Desert (below).

Rest stop
The China contingent take a break in Suzhou, a town in Gansu province southeast of the Gobi Desert.

Zarma warriors
Alexander Iacovleff (1887-1938), official artist for the Citroën expeditions to both Africa and Asia, produced this drawing in the Zarma lands of what is now western Niger (right).

horseback with his crew of scientists and film-makers in tow. The hostages were duly released, but only after the payment of a ransom consisting of three vehicles and two wireless transmitter-receivers.

Extreme conditions

As a result of the delays, the joint mission was forced to set off across southern Mongolia and the Gobi Desert in the depth of winter. Vehicles that had been designed with extreme heat in mind had to cope with temperatures as low as -30°C (-22°F). The engines had to be kept ticking over day and night to stop them freezing up. Mechanics found that their fingers stuck to the bare metal. Morale sank dangerously low. Yet despite such terrible conditions, the expedition – described by Audouin-Dubreuil as 'enthralling and

comradely, accursed and unhappy' – finally succeeded in reaching Beijing on 12 February, 1932, receiving a spectacular welcome from the city's diplomatic community.

Yet even then, the mission's troubles were far from over. On the return journey to France, by way of Indochina, India, Persia and Beirut, Haardt fell ill of pneumonia; he died in Hong Kong in March 1932. André Citroën sent the team a telegram: 'A man may be dead, but the work remains. Bring your leader's body back to France. My tears join yours.' In fact the mission was subsequently called off and eventually the team returned to France by sea.

'THE CAMEL IS DEAD'

The Central Asian Mission was not the first long-distance enterprise that André Citroën had dreamed up. Between October 1924 and June 1925 he dispatched eight half-track vehicles to make their way from Algeria to Madagascar, a journey of some 28,000km (17,500 miles). A camera crew accompanied the expedition and the film they produced, emphasising the enthusiastic reception of the vehicles and their technological wizardry, was shown at the Paris Opera in 1926 with the president of France in the audience. Never one to miss a publicity opportunity, Citroën proudly declared, 'The camel is dead, and Citroën cars have replaced it'.

Scientists on the road
Pierre Teilhard de Chardin (centre) lines up with other members of the China group. He had first travelled to China in 1923 on a mission for Paris's Museum of Natural History.

IN THE SERVICE OF SCIENCE

The Central Asian Mission included scientists in its ranks as well as mechanics. Pierre Teilhard de Chardin, a Jesuit priest and prominent paleontologist, used the opportunity to establish the first geological map of China. Joseph Hackin, curator of Paris's Musée Guimet, acquired objects of considerable value for his museum en route. Alexandre Iacovleff, a painter specialising in ethnographic subjects, was commissioned by the Ministry of Culture to record on canvas 'indigenous customs and ways of life that are gradually disappearing' and produced some extraordinary images of the peoples he met with along the way. The naturalist André Reymond brought back numerous specimens of China's flora and fauna. A film-maker André Sauvage shot many reels of film en route, although another director, Leon Poirier, was named in the credits when the documentary recording the trip finally came out in 1934.

An entrepreneur's sad end

As for André Citroën, the man who had put his firm's name up in lights on the Eiffel Tower in 1925, he barely had time to profit from the publicity surrounding the expedition. Having overstretched himself financially, he died a ruined man on 3 July, 1935. In the previous year his firm had introduced the revolutionary 7A front-wheel drive model, built to a monocoque design with bodywork and chassis arc-welded into a single unit – a car well in advance of anything else on the market at the time. The Michelin tyre company bought up the Citroën trademark, and the company went on to enjoy a prosperous future.

Other long-distance rallies and endurance tests have come and gone since André Citroën's day, notably the Paris-Dakar, linking France and Senegal, first staged in 1979. Yet the Central Asian Mission, conceived as a technological, scientific and cultural epic, still remains unparalleled as one of a kind.

Crossing the Himalayas
One of the half-tracks in the Pamir group struggles to traverse a road partly blocked by a landslide. The vehicles averaged about 12 miles a day.

Glass fibre 1931

In 1836 a Parisian named Igance Dubus-Bonnel took out a patent on a fabric made of glass, heated and drawn out in long fibres, and mixed with silk or linen. Making the material was difficult and costly, so only small quantities were produced at the time. In 1931 the invention was rediscovered by an American bottle-making firm, the Owens-Illinois Glass Company, which set out to find a market for the new fibre. Chevrolet chose the material – which was 30 per cent lighter than steel and easy to mould – for the bodywork of its new Corvette in 1953. Since then glass fibre has been used to make yacht hulls and Formula 1 racing cars, as well as in building insulation, reinforced concrete and composite materials. Since the 1960s it has been vital in the manufacture of optical fibres.

Production line
After the strands of glass have been extruded as fine fibres, they are mechanically spooled around spindles (above right).

MINERAL FIBRES

Glass wool and other similar substances such as slag wool (a by-product of the smelting process) are composed of bundles of mineral fibres produced when molten metal is sieved or spun in centrifuges. To achieve a stable and water-repellent result, a binder is added, along with oil to reduce dusting. The end product repels rot, parasites and rodents, making it useful as an insulating material. An added bonus is that mineral fibres are flameproof and can slow the spread of fires.

Car radios 1931

In-car entertainment
An elegantly dressed motorist of the 1930s tunes her car radio (left). The introduction of the Radio Data System (RDS) in the 1980s made it possible for long-distance travellers to listen to the same station throughout their journey without needing to retune.

The Galvin Manufacturing Corporation put the first car radio on the market in 1931 under the brand name Motorola. The product proved so successful that the company itself eventually adopted the name. Germany's Blaupunkt introduced the devices to the European market soon after, in 1932. Early car radios were cumbersome, weighing up to 15kg, and very expensive, costing as much as a third of the price of the car itself. They relied on the car battery for power and were usually installed in the boot; it was only in the late 1930s that they found a home in the dashboard. Twenty years later Blaupunkt produced the first FM models. Subsequent developments included the coming of stereo and preset stations and the addition of cassette players (in the 1960s), CD players (in the 1980s), then of Bluetooth and USB ports at the start of the 21st century.

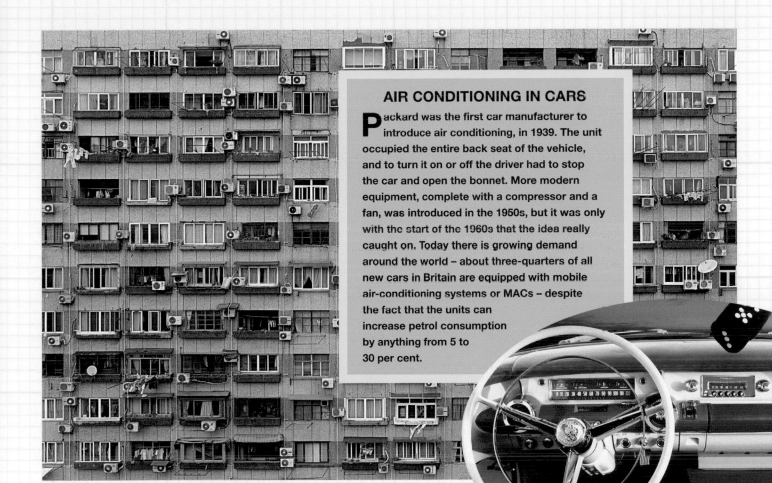

Air conditioning 1931

As early as 1902, Willis Haviland Carrier, a young engineer fresh out of Cornell University, developed an apparatus designed to maintain constant temperature and humidity in a printing works. Twenty-nine years later the Carrier Corporation that he founded sought to capitalise on his earlier work by unveiling the Atmospheric Cabinet, a device using ammonia as a liquid refrigerant to cool the temperature of a room. The Cabinet proved more effective than earlier systems, in use up to the end of the 19th century, that had aimed to cool air by passing it over ice. But there were problems with the design because ammonia was both toxic and flammable.

Reliable but expensive

Air conditioning became more reliable with the adoption of the gases trade-marked as Freons. These refrigerant chlorofluorocarbons (CFCs) were stable, fire-resistant and non-toxic. In the decades that followed, air conditioning gradually became a feature of theatres, cinemas, shops and offices across the USA. After the Second World War demand exploded, particularly in the southern Sun Belt, which attracted millions of new residents with the prospect of air-conditioned homes. Systems that heated air as well as cooling it were developed, as were devices that could filter air to provide a sterile environment for hospital operating theatres, or cleanrooms used in the production of high-tech electronic components such as integrated circuits. The downside of the spread of air conditioning was a surge in electricity use, while ageing or badly maintained units can be a breeding-ground for harmful micro-organisms.

AN EARLY PRECURSOR

Leonardo da Vinci foresaw the principle of air conditioning when he designed a paddlewheel device to divert fresh air from a watercourse to cool the apartments of his patron, Isabella d'Este.

Cool cars
Buick introduced air conditioning in its luxury convertibles in 1956 (above).

Chilling in China
Every apartment in this Shanghai tenement (top left) has its own air conditioner. The noisy outside units containing the compressors and fans to keep out warm air.

THE CONCOURS LÉPINE
An annual forum for inventions

Named after its founder, Louis Lépine, France's Concours Lépine is a unique showcase for amateur inventors. Entries in this competition for Heath Robinson-style gadgetry have included traffic lights for the colour blind, optoelectronic school clocks, spectacles for chickens and devices to help the elderly take off their socks without having to bend over.

THE COMPETITION'S FOUNDER

Besides being Prefect of Police for the Seine and Oise, which in his day covered all Paris, Louis Lépine was himself an innovator, seeking practical solutions for complicated problems. Noting that road junctions were becoming increasingly dangerous as automobile traffic grew, he introduced roundabouts across the French capital in 1906. To speed up the response of fire brigades to emergencies, he installed a citywide system of telephone alarms. He was also responsible for setting up a bicycle brigade and giving Paris its first river police.

Mug shot
Louis Lépine as e appeared on his anthropometric record card, in accordance with a system set up in the 1880s by Alphonse Bertillon to help to identify criminals and delinquents. The techniques introduced by Bertillon remained in use by police forces around the world until the 1970s.

When, in 1901, Paris's Prefect of Police Louis Lépine created the competition that bears his name, he was reflecting the growing prestige that inventing had come to enjoy over the previous century. In the Middle Ages most innovators were anonymous. As late as 1803 the *Gentleman's Magazine* had drawn up a list of 18th-century English luminaries without including a single inventor in their ranks. By Lépine's day, a century later, the situation was very different. The daily lives of ordinary people had been transformed through the contributions that new technologies had made to industrial progress in such fields as rail travel, domestic heating, electric light, telephones and cars, to name just a few.

Heroes of a new age

Feted at a succession of exhibitions and World's Fairs, 19th-century inventors had found themselves covered in unexpected glory.

In 1862 some 70,000 people attended the unveiling of a statue of George Stephenson, inventor of the locomotive, in Newcastle upon Tyne. By that time innovation had come to be seen as critical for the creation of jobs and businesses, as symbolised by Thomas Edison, the man of a thousand patents and founder of General Electric.

Encouraging creativity

A prolific innovator himself, Lépine started the competition at a time when the small businesses of Paris were feeling the effects of an economic recession made worse by competition from low-cost manufacturers abroad. The first prize-winning entry was a construction set along the lines of Meccano. Then, as now, only patented inventions could be entered, and they had to be both useful and economically viable – in other words, suitable for mass-market production and distribution.

The point of the competition has never really been the financial rewards, which are little more than symbolic. In 1901 the first prize, grandly entitled the Prize of the President of the Republic, was worth just

100 francs, and even today amounts to only 3,000 euros (about £2,500). The total sum distributed in 2009 came to about 30,500 euros (£25,000), shared between some 650 different winners. Of far greater importance is the fact of being selected for the competition, which in itself attracts the attention of entrepreneurs.

Freelances of the imagination

Over the years the competition, which for a long time remained unique in the world, has attracted a good number of foreign entrants. For the most part the contestants have paid

Prototype pump
Designed by M Henry, this artificial heart (below) was entered in the 1937 competition. Three years earlier, the same inventor had exhibited a blood transfusion mechanism.

Flight of fancy
A hopeful entrant makes his way to the Concours in 1935 (above). Today, in addition to the main international competition held in Paris, whose top award is the Prize of the President of the Republic, there is a Europe-wide contest based in Strasbourg as well as some regional events.

Snail trap
Put on show in 1950, this device was designed to catch unwary molluscs.

out of their own pockets to develop the devices they put on display and the inventions cover every imaginable field. Many are aimed at the general public, including games, household gadgets and aids for the elderly or the handicapped. Other inventors target industry, whether in the form of new products or improved production methods for existing ones. The result is an eccentric mix that helps to explain the perennial fascination that the competition holds for the French media.

So who are the people responsible for so much inventive creativity? They come from all walks of life and most work on their own. Some are old, a few very young – in 2005 two 14-year-olds won a prize for a magic letter-box. Very few are professional engineers.

Addressing domestic concerns *Household appliances presented to the judges at the Concours have included an electric tie iron (left), a coffee grinder designed by the Peugeot brothers, more famous as car manufacturers (centre left), and the vegetable mill created by Jean Mantelet, founder of the Moulinex cookware brand (bottom left).*

Ingenious notions

Most of the ideas aimed at the general public address practical needs, such as reducing water or electricity consumption, speeding up cooking times or saving space in crowded apartments. Life's minor inconveniences can trigger flashes of inspiration: magnetic window-washers that clean both sides of the glass at the same time, for example, or umbrellas that will not turn inside out. There was the handbag that lit up inside when opened (entered in 2004) and the triple-bladed slicer (2009).

Other inventions seek to improve on products that are already in widespread use, addressing some drawback overlooked in the original design. Examples include the CD case

TOO FEW WOMEN WINNERS

Since its inception in 1901, only three women have won the top prize in the Concours Lépine, although it has been suggested that some of the household appliances submitted might have actually been invented by women but entered in their husbands' names. The first female winner was a Mme Labrousse, who entered a design for a washing machine in 1912. The other two came in the present century, in 2001 and 2009 respectively.

with a push-button opener from the 2000 competition, the headphone wire rewinder presented the following year, or the accessory designed to clip skis and ski-poles together displayed in 2009. Some focus on industrial problems, like the robot window-cleaner entered in 2000 or the anti-friction revolving joint for gas-powered machinery from 2009.

Some of the gadgets can seem ridiculous at first sight, such as the 'chicken spectacles' that an American inventor named Irvin Wise presented outside the exhibition in 1960, or the cat-sack designed to get rid of fleas. Yet even these had a serious purpose: restricting chickens' vision can reduce aggression in the coup, for instance, while a cat enveloped within the sack would thoroughly impregnate itself with flea powder.

Ideas are not enough

Inventions have been known to reappear in different forms over the years – accident-proof devices for opening oysters, for example, have turned up six times in half a century. Designs for hands-free umbrellas have been popular, most of them strapped to the user's head or shoulders. The reason why an idea comes up again and again is not necessarily because previous gadgets fail to solve the particular problem. In many cases the difficulty is that the creators get so carried away by the ingenious nature of their inventions that they neglect the importance of marketing. Commercialising a new product can take more effort and resources than making it in the first place. The Hungarian inventor László Bíró, for example, sold very few of the pens that bear his name before Baron Marcel Bich, the co-founder of Bic pens, bought out his patent.

The same could be said of Étienne Mollier, who won the gold medal at Concours Lépine in 1910 for a camera that used perforated 35mm cine-film to take 100 shots in succession. Mollier had little success in marketing his idea, which was completely novel at the time, yet 15 years later Leica, the German camera company, made 35mm film the reel of choice for its standard 24 x 36mm format cameras. In 1952 a bronze medal went to Roger Dambron for a game that involved assembling different parts of photographs to form a whole, but it took a police inspector to find a use for the invention by developing photofit images.

The Concours has had its commercial successes, although not many of them. The Bac Riviera humidifier, designed to keep plants in window boxes watered, has become a standard feature on many French balconies, although it remains relatively little known in

Spectacles for chickens
These little plastic devices (right), equipped with pins, could be fixed to chickens' beaks as blinkers, deterring the birds from pecking one another.

Britain. A road-race card game called *Mille Bornes* ('A Thousand Milestones') was distributed for a time by Parker Bros in the USA and is still available there. And a product known as Aspivenin, which won the 1983 gold medal, is now widely available. A miniature suction pump designed to extract insect stings, it has become required equipment for summer-camp supervisors and finds its way into many walkers' rucksacks.

A repository of ideas

Nowadays more than half the inventions accepted for the competition subsequently find their way onto the market, whether put there by the inventors themselves or else by some firm to whom they have sold the patent rights. The figure is remarkably high, but the effect on the general public has not been especially great. Most of the devices involved tend to be widgets of limited use aimed at niche markets. Even so, they perfectly fulfil Louis Lépine's original ambition: to encourage small and medium-size enterprise.

Competitors are sensitive to the concerns of the day, as suggested by the plastic bottle crusher presented in 2005, the energy-efficient drying

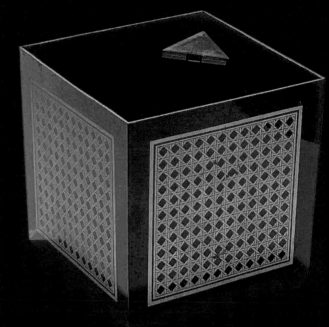

Humidifier for plants
The Bac Riviera plant holder (right) has a reservoir in its base from which water rises upward through capillary action. The system provides constant humidity.

Extracting the sting
Acting like a reverse syringe, the Aspivenin pump (below) creates a vacuum that draws out snake venom or a jellyfish sting.

KIT de premier secours
THE EXTRACTOR
LA PLUS EFFICACE
MÉTHODE VALABLE
pour le traitement des
piqûres et morsures en
première urgence

Comment
utiliser
l'Extractor

SERPENTS   ABEILLES   AUTRES

THE EXTRACTOR

A SURREAL TOUCH

Arthur Paul Pedrick was a prolific British patentor of inventions. He took out 161 patents between 1962 – the year of his retirement – and 1976, when he died. But he never sought to capitalise on his ideas, which tended towards the surreal. One of his plans was to transport snowballs from Antarctica to Australia by pipeline, another was to divert water from the Amazon Basin to the Sahara Desert. A former employee of the Patent Bureau, his main aim was in fact to point out inadequacies in existing patent law.

Building a photofit image
The photofit system allows police to fit together facial features from different individuals to create a composite image matching descriptions of a suspect gathered from witnesses.

cupboard using a refrigerant dehumidifier in 2000 and the drone designed to detect chemical and radioactive pollution in 2007.

A forum on the internet?

The past few decades have seen various fresh initiatives to encourage invention, among them the International Salon of Inventions set up in Geneva in 1972 and the reality television show *Dragon's Den* in which inventors pitch for investment. Launched in Britain in 2005, the programme started in Japan in 2001 and now has versions in Israel, Australia and even Afghanistan. Yet many of the 80,000 patents taken out around the world each year are never exploited for want of entrepreneurs willing to work with the patent-holders to develop the ideas commercially.

Might the internet make a difference? There are websites devoted to unexploited patents, and the monitoring of technological developments that all big firms undertake ensures that these are trawled for possibilities. Multinational corporations engaged in high-tech fields such as electronics might have as many as 50,000 patents in their portfolio at any one time, taking out hundreds each year. Investment often pays off handsomely, a fact learned by Britain's Trevor Baylis, inventor of the wind-up radio. Because he could not afford to patent many of the inventions that preceded the radio, he was inspired to found the Trevor Baylis Foundation to promote, encourage and support inventors and engineers.

Solar home
The Heliodrome (above) designed by Eric Wasser won first prize in the 2003 competition. Its curved form allows it to take maximum advantage of winter sunlight.

Intelligent clothing
Presented in the year 2000, this Olivier Lapidus blouse featured a telephone as an integral part of one of its sleeves. As such, it prefigured the work now being done on e-textiles that take advantage of recent advances in IT and nanotechnology.

PROLIFIC INVENTORS

The competitor who has won the most prizes at the Concours Lépine is Claude Dumas, with a total of 29 medals over the years. Born in 1936, Dumas has chalked up 214 inventions and innovations, most of them made of wood, from a revolving CD rack to a small folding table that expands to seat 18 dinner guests. Yet the cost of taking out patents is such that some of his ideas have gone unprotected and been copied by others. The man who claims to be the world's most prolific inventor is the eccentric Japanese Dr NakaMats who has succeeded in being granted more than 3,300 patents for items as diverse as floppy discs and karaoke machines. Among his least successful inventions are an aphrodisiac called the Love Jet and an intellect-improving Brain Drink.

Supernovas

1934

Supernovas form a special category of exceptionally bright stars. Readers of the *Proceedings of the US National Academy of Sciences* could have learned as much from a report by Walter Baade and Fritz Zwicky published in the March 1934 issue. Progress in astrophysics allowed the two astronomers to coin the word itself (which they hyphenated as 'super-nova'), as well as to make the claim that the phenomena were actually stellar explosions marking the death of stars.

Baade and Zwicky also established that supernovas come in two types. The first of these, thermonuclear supernovas, we now know are produced in binary star systems (in which two stars jointly orbit around one another), when the older of the two stars – a white dwarf – begins to draw off the hydrogen envelope surrounding its companion. After slowly accumulating on the white dwarf's surface, the gas abruptly undergoes nuclear fusion, causing the star to explode.

The second type results from gravitational collapse. This occurs only in the most massive stars – those with a mass at least eight times greater than that of our own Sun. Such stars spend most of their active life transforming hydrogen into helium through fusion reactions. The outward pressure that the radiation creates is balanced by the gravitational forces holding the atoms together, maintaining the star in a state of equilibrium. But when the star's core begins to run out of hydrogen, gravity reasserts itself, producing a rise in temperature. This causes the helium in its turn to enter a state of fusion, so the star continues to shine. Over successive cycles, all the elements of the stellar core end up being transformed into iron, which is not susceptible to fusion. At that point, the star's core metamorphoses into a neutron star or even a black hole, while the blast wave caused by this final explosion sweeps out stellar debris, creating a nebula – an immense cloud of interstellar gas and dust.

The brightness of supernovas is prodigious. At its peak the intensity of the light given off by the dying star can equal that of the entire galaxy to which it belongs.

SN1987A: SURPRISE IN THE SKIES

Supernovas occur on average between one and four times a century – so rarely, in fact, that the appearance of one in 1987 just 2.5 million light years from Earth was a welcome surprise for astronomers. Observation of the exploding star, labelled SN1987A, confirmed that most of its energy was given off in the form of neutrinos – particles with very small mass – and that the blast wave released by the explosion is gradually forming a coloured nebula. The composite image below was taken from the Hubble Space Telescope in November 2003.

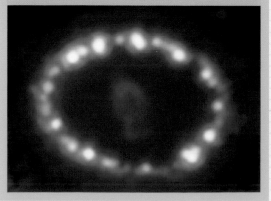

Death of a star
The Crab Nebula in the constellation of Taurus (above) is made up of the remnants of a supernova observed by astronomers in China between AD 1054 and 1056. Brighter than all the stars in the sky, it remained visible in broad daylight for several weeks.

Pinball 1934

A s long ago as the 17th century people played a sort of mini-billiards known as bagatelle. This involved using a short wooden cue to launch a small ball onto a playing area dotted with holes, each worth a set number of points. Much later, in 1932, there was a vogue for so-called 'pin games' such as Baffle Ball, devised by the American David Gottlieb, and Ballyhoo invented by Raymond T Moloney, who went on to found the Bally games company in Chicago. Both of these used a spring-loaded plunger to fire the ball.

Players soon learned to 'nudge' the boards to steer the ball in the direction they wanted. In response, in 1934 a game designer named Harry Williams took advantage of the recent availability of electrically powered solenoids to devise a tilt mechanism that automatically stopped the game if the manipulation was too brutal. By doing so he transformed pin games into pinball. Bally subsequently introduced mushroom-like bumpers for the ball to bounce off. Flippers followed in 1947 when Gottlieb introduced his Humpty Dumpty model.

Lighting up
A pinball game with a Moon-landing theme designed in 1990 by the American photorealist painter Charles Bell.

The Richter scale 1935

Measuring catastrophe *Like all the original Richter-scale readings, this one (below left) was collected by a seismograph located 100km (60 miles) from the centre of the quake.*

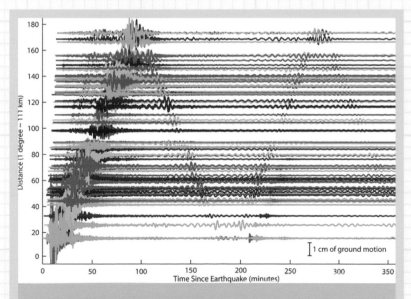

THE INDIAN OCEAN EARTHQUAKE

J ust before 8.00am on 26 December, 2004, an earthquake with a magnitude measuring between 9.1 and 9.3 on the Richter scale occurred off the coast of the Indonesian island of Sumatra. One of the most violent quakes ever recorded, it unleashed a tsunami that devastated not just Indonesia and other Southeast Asian nations but also Sri Lanka and southern India. The known human toll eventually reached 222,046 people dead or missing.

I n 1935 the American geologist Charles Richter devised a way of calculating the local magnitude of earthquakes by measuring the amplitude of the largest seismic wave at ground level. The scale that he drew up increased in single units from 1; theoretically the scale has no top limit, but in practice the worst quakes so far recorded have registered between 9 and 10. The sequence is logarithmic, which means that each successive number represents a force 10 times greater than the one below it. So, for instance, a quake measuring 6 on the Richter scale is 10 times more powerful than one rated 5.

In 1956, in collaboration with Beno Gutenberg, Richter came up with a more precise model, calculated on the basis of the amplitude of the shockwaves that form deep beneath the Earth's surface when a quake occurs. It was this second measure devised by Richter that replaced the earlier Mercalli scale.

In 1977 a Japanese seismologist, Hiroo Kanamori, developed the moment magnitude scale, considered the most accurate scale of all. It allows scientists to quantify the energy released by a quake at source, taking account of fissures and earth movements, as well as the heat released and the seismic waves.

Photography brightens up

The coming of Kodachrome film, launched in April 1935, brought colour photography within the reach of all pockets. It was the brainchild of two friends, Leopold Godowsky and Leopold Mannes.

The two Leopolds were still at college in America when they decided to invent a film that could capture images in colour. For many years they worked on the idea in their spare time, without success, while building careers as classical musicians. Then, in 1930, Eastman Kodak agreed to provide the pair with laboratory space to advance their work.

Faithful reproductions

Godowsky and Mannes had experimented with additive colour mixing, attaching colour filters to cameras and projectors employing black-and-white film. Now, they came up with the idea of a film composed of three separate layers, the top one sensitive only to blue, the middle one to blue and green, and the bottom one to blue and red. No colouring agent was involved. Instead, the chemical that brought out the pigmentation was contained in the developer rather than in the film itself, as it had been in their earlier attempts when it had the irritating effect of mixing up the colours. This time the results were striking. The new film reconstituted colours in the right place and in the correct proportions. One disadvantage was that because the film was susceptible to every nuance of shade, it had to be developed in total darkness.

Photo opportunity
Visitors to the New York World's Fair of 1939 (above) pose in front of an Eastman Kodak publicity backdrop. Based in Rochester, New York, the firm played a central role in making photography a popular hobby, launching the mass-market Brownie camera in 1900.

COLOUR PHOTOGRAPHY BEFORE COLOUR FILM

In 1855 the Scottish physicist James Clerk Maxwell had the idea of creating colour photographs by using separate red, green and blue filters, then projecting the three negatives simultaneously onto a screen, using filters of the same colours on each of the projectors. Using Maxwell's 'formula', Thomas Sutton made the first colour photographic image, of a tartan ribbon, in 1861. Eight years later two French inventors, Charles Cros and Louis Ducos du Hauron, separately came up with the idea of so-called 'subtractive' colour through a process that involved taking three separate negatives, then transferring the positive images onto plates coated with gelatin containing differently coloured pigments; these images were carefully superposed on top of one another. In 1907 the Lumière brothers started selling heavy, fragile glass plates called *autochromes Lumière*, each one covered in a film of microscopic grains of dyed potato starch that acted as colour filters. The system only produced single images, not prints or copies, which proved a fatal drawback in the long term.

Historic projector *Made in 1900 for the chemist and photographic pioneer William Abney, this device employed James Clerk Maxwell's filter system.*

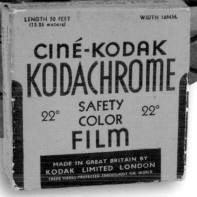

THE FIRST COMPACT CAMERA

A German engineer named Oskar Barnack, growing tired of lugging around heavy photographic equipment, developed the Leica 1 (below), which was unveiled at the Leipzig Spring Fair in 1925. It was the first compact camera and it pioneered the 35mm film format. It was not an immediate success, looking too much like a toy for popular taste, but users quickly learned to appreciate its exceptional optical qualities, along with the fact that it was both hard-wearing and easy to operate. Before too long the Leica 1 was a favourite with photojournalists and amateurs alike, setting off something of a revolution in the world of photography.

PHOTOGENIC RECORD
In time the word Kodachrome became so familiar that a national park in Utah now bears the name of the Kodachrome Basin State Park.

Landmark products
Kodak's best-known products included Kodachrome colour film (above), first launched in 1935, and the slides developed from it (below).

A world of colour

Kodachrome film was first launched in 16mm movie form, then a year later was made available in 35mm slide format. The German firm Agfa soon brought out a competing product. Colour film quickly proved a success with the public, who appreciated its strong contrasts and colour retention, as well as the fact that it was simple to use. Despite competition from Kodacolor, which provided users with prints rather than transparencies, Kodachrome retained the loyalty of millions of amateur photographers who used it to take holiday snaps and to record their children's first faltering steps for posterity. Friends and family would happily gather in the darkness to watch the results, with each new image heralded by the clicking of the projector. Many of the best-known news and fashion photographers also became enthusiastic users of the film.

Yet Kodachrome could not withstand the coming of digital photography around the turn of the new millennium. Kodak stopped producing slide projectors in 2004, and five years later withdrew Kodachrome itself, a decision that symbolised the passing of the age of analogue imagery.

Cortisone 1935

In the early 1930s an American chemistry researcher named Edward Kendall, working for the Mayo Clinic in Rochester, New York, set out to discover the secrets of the adrenal gland. Working with samples from cattle, his team succeeded in isolating 10 different compounds, each designated by a letter of the alphabet. Compound E, identified in 1935, would later become known as cortisone. Kendall immediately realised that it might provide a treatment for Addison's disease (an endocrine disorder caused by adrenal insufficiency), and also thought it could prove useful in cases of shock, burns and stress reactions, among other conditions.

One of Kendall's colleagues, a doctor called Philip Hench, foresaw other possible uses. He had noticed that the pain associated with rheumatoid arthritis was alleviated in women who were pregnant. Reckoning this was due to a hormonal reaction, he put forward the idea that hormones could themselves be used as medicines. In 1948 he and Kendall gave Compound E to a young woman suffering from severe arthritis. The results were immediate and spectacular. They subsequently extended the trial to 13 patients treated over a six-month period, with such success that several different pharmaceutical firms started synthesising the compound. The competition encouraged the development of cortisone derivatives that were to prove even more powerful in their effects.

Despite some unwanted and unpleasant side-effects associated with long-term use, including weight gain and oedemas (swellings), corticosteroids came to be employed for the treatment not just of rheumatism but also for asthma, allergies and other conditions. Today they continue to be the standard treatment for many inflammatory ailments.

Model molecule
In this 3D image of a cortisol molecule, carbon atoms are shown in blue, hydrogen in yellow and oxygen in red.

Chemist at work
Edward Kendall photographed in his laboratory. His work on cortisone won him the 1950 Nobel prize for medicine, in company with his collaborator Philip Hench and the Swiss chemist Tadeusz Reichstein.

THE STRESS HORMONE

During the Second World War, the American intelligence services received word that Nazi agents were collecting bovine adrenal glands in Argentina to protect Luftwaffe pilots from combat stress. After hearing this report, synthesising cortisone for a time became a US priority. It later turned out that the CIA had been misinformed, although the hormone is indeed used to treat stress.

Opinion polls 1935

In 1935 the US publicist George H Gallup founded the American Institute of Public Opinion. There was nothing inherently new about the idea of investigating the popular mood. In 1745, for example, Louis XV of France had encouraged government officials to spread rumours in order to judge citizens' reactions to the news. As early as the 1824 presidential election in the USA, American newspapers had organised 'straw polls', printing reply coupons for their readers to indicate the way in which they intended to vote. But what was new about Gallup's 1935 venture was the idea of polling a representative sample of the population as a whole.

The presidential election the following year, 1936, demonstrated the validity of Gallup's new approach. A straw poll conducted by the *Literary Digest* drew more than 2 million responses and yet it wrongly indicated that the Republican candidate, Alf Landon, would win. Gallup, with a sample of just 5,000 voters, correctly predicted victory for President Roosevelt. Despite a hiccup in 1948, when the organisation wrongly forecast victory for Thomas Dewey over the Democrat incumbent Harry S Truman, the polls quickly became an established national institution.

By the 1940s Gallup had set up a British branch of his organisation and this correctly foresaw the Labour Party's victory over Churchill's Conservatives in the 1945 general election. The rival MORI (Market & Opinion Research International) polling organisation was set up by Robert Worcester in 1969. From the 1970s on, opinion polls have become an essential tool of both marketing and politics.

SAMPLING

The accuracy of any poll depends largely on the way in which the questions are expressed and the size and quality of the sample employed. This can be chosen randomly, on the principle that every element of the general public has an equal chance of being included, or according to a system of quotas designed to represent the 'profile' of the targeted population.

Inaccurate opinion *Newly re-elected President Truman jubilantly holds up the* Chicago Daily Tribune *showing a headline wrongly reporting his defeat in November 1948.*

The parking meter 1935

Parking on parade *A line up of cars and meters in Omaha in 1938. Unlike meters, which are specific to a single parking space, the more recent pay-and-display ticket machines service multiple spaces simultaneously.*

Given the job of supervising car parking in Oklahoma City, Carlton Magee, an employee of the local chamber of commerce, had the idea of installing machines that would collect money in exchange for the right to park in a given space for a fixed time. The meters he devised contained a clock mechanism that started counting down as soon as coins were fed through the slot; meanwhile a moving flag indicated the paid-for parking time remaining. Besides ensuring a rotation of vehicles, the devices proved to be substantial money-earners for the city. Magee founded the Magee–Hale Park-O-Meter Company to manufacture the machines, the initial 150 of which were installed on Oklahoma City's streets in July 1935. Britain's first meters were arrived in Grosvenor Square, London, in 1958.

Using radio waves to see the unseen

Radar was developed in Britain in the 1930s for the purposes of aerial defence. Since that time it has been put to many non-military uses. It works by detecting reflected radio waves.

Defensive beacon
This Chain Home beacon (right) was one of 60 such towers making up the system that defended Britain's coasts in the Second World War. The pylons supported an array of transmitting antennas, while the return signals were picked up via slightly smaller wooden receiving towers.

Far-sighted Britons had good reason to view the skies with alarm in the mid 1930s. Germany was rebuilding its armed forces in direct contravention of the Treaty of Versailles, and its factories were turning out bombers at a steady pace. The question was, if war did break out, how best to deal with planes that could cross the Channel in a matter of minutes. Most people thought that the only recourse lay in anti-aircraft defences and the skills of RAF pilots. Winston Churchill did not share that view, believing that science could be called upon to help defend the country.

Safer skies

At the time the future war leader was in his wilderness years as a backbench MP, but he was nonetheless able to engineer the selection of a sympathetic scientist onto the Aeronautical Research Committee (ARC), chaired by Sir Henry Tizard. The committee was increasingly concerned with Britain's aerial defences, and its work proved highly influential.

With the ARC's encouragement, in February 1935 Robert Watson-Watt of the National Physical Laboratory sent a secret memorandum entitled *Detection and Location of Aircraft by Radio Methods* to the Air Ministry. The document is often considered to mark the birth of modern radar, followed as it was by a successful demonstration of the new technology to the committee twelve days later. The result was that from 1937 Britain was provided with a string of radar stations, each with a range of 100 miles (160km), in a system codenamed Chain Home.

The telemobiloscope

Watson-Watt's work was hugely important, but he could not claim to have invented radar single-handed. In fact it was the end product of four decades of research and development. In 1886 Heinrich Hertz had noted that there were no fundamental differences between light

waves and electromagnetic waves. Fourteen years later, Serbian-born US scientist Nikola Tesla explored the possibility of pinpointing an object in motion with the aid of echoes from radio waves – in effect the guiding principle of radar, which detects objects that would otherwise be invisible by 'bouncing' radio waves off them and capturing the reflection.

Christian Hülsmeyer, a 23-year-old German, demonstrated the validity of Tesla's notion in 1904 by inventing a device called the telemobiloscope. Designed to prevent collisions between ships, the machine emitted radio waves that rebounded on coming into contact with a metal hull, returning back to

Within the web
Operators work a US Navy device during World War II. The American military developed their own radar technology independent of Britain in the course of the 1930s. Their CXAM system was tested in 1938 and went into production for the US Navy from 1940.

A 'DEGENERATE' SCIENCE

In 1935 Britain and Germany were neck-and-neck in terms of radar development, but by 1940 Hitler had withdrawn support for what he considered to be a defensive and therefore degenerate branch of research.

their source. The signals were captured by an antenna attached to a receiver, thus revealing the presence of a vessel lying in the direction in which the device was pointing. Hülsmeyer demonstrated his invention on the Rhine in May 1904, but at the time – eight years before the *Titanic* disaster made people more aware of the need for navigational security – no-one showed much interest in following it up.

Between 1922 and 1927 various researchers hesitantly retraced the route that Hülsmeyer had traversed two decades earlier. In September 1922 two Americans, Albert Taylor and Leo Young, were conducting radio experiments across the Potomac River near Washington DC when they noticed that passing boats interfered with the signal. The two, who were working for the US Navy at the time, duly put forward the idea that radio waves could be used to detect enemy structures at night or in fog. In January 1930 one of their colleagues, Lawrence A Hyland, located a passing plane in similar fashion. The US Naval Research Laboratory thereupon decided to put more systematic effort into research on the detection of aircraft with the aid of radio waves.

Radar goes to war

Even so, Watson-Watt's work in 1935 marked a turning-point for radar. In addition to brilliantly synthesising the work previously done in the field, he championed the use of pulse radar in place of the continuous waves

CAPE MATAPAN: A VICTORY FOR RADAR

On 26 March, 1941, Italian Admiral Angelo Iachino massed his fleet off Crete with the intention of ambushing vessels of the British Navy. Two days later the two fleets clashed, but the encounter proved indecisive and when night fell the Italian squadron retired toward its base. Thanks to the radar installed on a couple of his vessels, the British commander Admiral Andrew Cunningham learned that two enemy cruisers were heading straight for his fleet, hoping to take advantage of the darkness to go to the assistance of one of their own ships that had got into difficulties. At 10.28pm precisely, the destroyer HMS *Greyhound* abruptly switched on its searchlights, illuminating the cruiser *Fiume* at short range and taking the Italians completely by surprise. British ships opened fire and within half an hour the *Fiume* was sunk. The Italians lost three cruisers and two destroyers in the battle, while the British squadron returned to port without loss. Radar's usefulness in night fighting had been amply demonstrated.

All-seeing eye
This new 'seeing-eye' radar equipment was successfully tested in 1949. The device offered the advantage of a wide signalling field, considerably improving aerial detection rates.

Spy in the sky
An AWACS plane – the acronym stands for Advanced Warning and Control System – from the NATO fleet. The plane is a Boeing 707 converted for military use by the addition of the disc-shaped radome that scans the skies for low-flying aircraft.

used up to that time. One advantage was that the pulses made it possible to calculate the range of the target, by measuring the interval of time between the original transmission of the signal and its return. In 1936 Watson-Watt made the system more effective by improving the focusing of the radar beam.

By that time modern radar was effectively up and running, although it still lacked a name. That was supplied by the US Navy in 1940, when 'radar' – an acronym for Radio Detection and Ranging – was adopted.

Radar won its spurs as an essential mechanism of Allied defence in the course of the Second World War. The Chain Home stations in particular played a decisive role in the Battle of Britain. On 12 August, 1940, for example, they helped to thwart a Luftwaffe attack on the factories along the River Tyne, an attack that would probably have succeeded in doing serious damage to Britain's war effort if local radar stations had not given the RAF almost an hour's warning in which to prepare.

USING RADAR TO FORECAST WEATHER

In the 1940s radar operators learned by chance that rain, snow and hail all reflected electromagnetic waves. Britain's Meteorological Office began experimenting with radar in the 1950s, but many years would elapse before the system was sufficiently sophisticated to predict rainfall at a local level. Today the Met Office has 18 radar stations around the UK and Ireland, each with a range of about 150 miles (250km). After initial analysis in the stations' own computers, the data is transferred to Met Office HQ in Exeter to build a national picture. While accurate for steady rainfall, the system struggles to cope with drizzle if the droplets are too small to register.

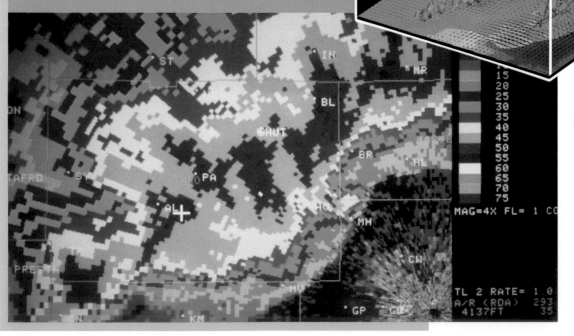

Plotting rainfall
Thanks to the Doppler effect, radar can detect how fast rain showers are moving; the different colours on the screen (left) correspond to varying amounts of precipitation. The computerised atmospheric model (above), one of a number introduced in the USA and Europe in recent years, uses numerical analysis of radar signals to give accurate local forecasts.

MEASURING SEA LEVELS FROM SPACE

Since the launch of the European Space Agency's ERS-1 satellite in 1991, radar altimeters have been used to measure the depth of the oceans from space. The information sent back is used in drawing up weather bulletins and preparing navigational maps, as well as for real-time surveillance of the ocean's surface, for instance to give warning of tsunamis. The satellite also helps to monitor global warming, which can be seen in a rise in ocean levels. The Jason-2, a joint US–French venture launched in 2008, can now measure sea level to within 2cm within 15 miles (24km) of the coast.

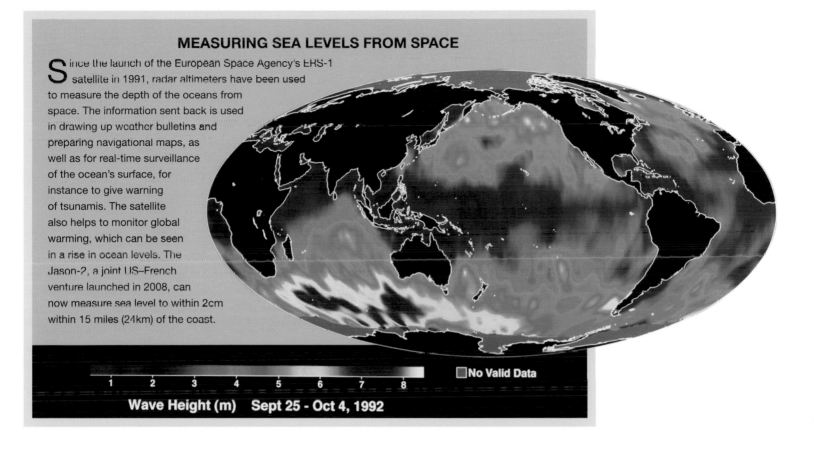

☐ No Valid Data

| 1 | 2 | 3 | 4 | 5 | 6 | 7 | 8 |

Wave Height (m)   Sept 25 - Oct 4, 1992

Relief map *This colour topographic chart of the oceans (above) was drawn up from information provided by the joint US–French TOPEX/Poseidon satellite, replaced in 2008 by Jason-2.*

An indispensable tool

After the war the range and accuracy of the system continued to be improved. The development of numerical analysis was probably the most significant innovation: first applied to electronic signals by the American statisticians John Tukey and J W Cooley in 1965, this was greatly enhanced by the later development of computers, making it possible to identify echoes and to clean unwanted interference and clutter from radar images.

Today radar plays many roles. It is still essential for military surveillance, whether via ground-to-air stations or equipment installed in aircraft or on boats, providing forces with a permanent three-dimensional picture of the air space around them. In addition, air traffic controllers rely on radar to manage the skies, radar speed guns are now used to catch unwary motorists, while weather forecasters employ it to spot precipitation. Radar also provides anti-collision systems on boats that call to mind Hülsmeyer's original telemobiloscope, which can now be seen as a pioneering invention well ahead of its time.

SPEED GUNS

The roadside radar guns used by police are not intended to detect the presence of cars so much as to measure their speed. They do so thanks to the Doppler effect, which causes an echo bounced off the object in motion to return to the radar receiver with a modified frequency and phase. Calculating the difference between the wave as initially transmitted and as it bounces back provides a reading on the speed of the target. Researchers had known the science of the Doppler effect on radio waves since the 1930s, but it was only 30 years later that electronics advanced sufficiently to provide reliable measuring devices.

Measuring speed with hand-held radar
Today's radar guns use a laser beam to measure the target's range instantaneously, showing the speed on an LED display.

LEISURE TIME
Recreation for all

In 1938 the British Parliament passed the Holidays with Pay Act, guaranteeing workers a week's paid leave each year. The act supplemented legislation of 1871 that established bank holidays as days of rest. The trend was an international one – across the industrialised world, more people than ever before could get out and about to enjoy their leisure time.

One horse power
Holiday-makers enjoy a sunny day on the Shropshire Union Canal in 1939. Once the vital arteries of industrial Britain, the canals were increasingly turned over to recreational use in the 20th century.

Britain was far from setting the pace in introducing paid vacations. Many countries had preceded it including Germany in 1905, Denmark and Norway in 1910, Finland, Italy, Czechoslovakia and Poland between 1919 and 1925, and Greece, Spain, Sweden and Brazil between 1926 and the mid 1930s.

Yet holidays were already very much part of the British mindset. In particular, the idea of a day out at the seaside had become part of the national psyche. In the case of paid holidays, the 1938 Act did not in itself create the practice; rather, like most social legislation, it made a legal right of a trend that was already well established. By 1925 around 1.5 million employees in Britain already had paid holidays as a result of deals negotiated between trade unions and employers; by 1937 that figure had risen to 4 million. Yet the Holidays with Pay Act did make a real difference: the equivalent number for 1939 was 11 million workers.

Even so, working-class families had to scrimp and save to take advantage of their new rights. By the late 1930s a week's holiday was estimated to cost a family of four £10, at a time when average weekly earnings amounted to just £3 10 shillings (£3.50).

Beside the seaside

Yet people managed. Around Britain's coasts, resorts that had risen to prominence in late Victorian times now expanded to meet the new demand. Brighton, Eastbourne and Bognor on the south coast, Ramsgate and Margate in Kent, Southend and Clacton in Essex all served holiday-makers from London and the home counties. In Yorkshire, workers in Leeds, Bradford and Sheffield flocked to Scarborough, Filey and Bridlington on the east coast, while employees in the Lancashire mill towns opted for Blackpool or Morecambe or the resorts of North Wales. In terms of sheer visitor numbers, Blackpool was unarguably the queen of the

A TASTE FOR THE SUN

In the 1920s the preference for pale skin, considered a mark of gentility among Europe's upper classes for centuries past, began to give way to a new appreciation of the sun. With the growing popularity of seaside holidays, a tanned skin came to be seen as a sign of health, wealth and vitality. French fashion designer Coco Chanel helped to set the trend by taking to the beach to sunbathe. There were strict limits, though, on how much skin could be exposed in public. In the interwar years, Brighton was the only resort in Britain that permitted men to bare their chests; elsewhere, they remained covered by singlets. Women wore one-piece bathing suits, although these were becoming more daring, with narrow straps and scooped-out designs that exposed much of the back.

Soaking up the sun
The new taste for suntans created a demand for protective sun lotions. The first sunscreen is said to have been created in Austria by Franz Greiter, founder of the Piz Buin brand.

resorts. One sunny Bank Holiday Monday in 1937, it attracted more than half a million day-trippers. By that time the town was pulling in more than 7 million overnight guests a year. The sun was the star draw, but alas it was not always forthcoming. So new attractions had to be built to keep visitors happy even in poor weather when the beach was not much fun: cinemas, funfairs, bandstands and dance halls all sprang up. Blackpool's famous Illuminations, first tried out experimentally before the First World War, became an annual feature after 1925, extending the season into autumn. At about the same time, the Pleasure Beach amusement park moved to its present site.

TOTALITARIAN FUN

In the 1930s Fascist Italy and Nazi Germany both set up holiday organisations that promoted their new ideologies. Italy's *Dopolavoro* ('After Work') organised sports and cultural activities that aimed to 'elevate physically, intellectually and morally'. The Nazis' *Kraft durch Freude* ('Strength through Joy') programme was a large-scale operation that arranged, among other things, cruises on which managers and workers could rub shoulders. Socially beneficial vacations became a moral duty.

Party-sponsored pleasure
German holiday-makers on a 'Strength through Joy' cruise wave to well-wishers as their ship sets sail. Large vessels like the 25,000-tonne Wilhelm Gustloff *were specially commissioned by the organisation, which inculcated Nazi values through its holidays.*

The great outdoors
Camping became increasingly popular in Britain between the wars. 'Pioneer camps' offering holidays under canvas were introduced in the 1920s.

Skiing in the sun
Only wealthy Britons could afford holidays at places like the French Alpine resort of Chamonix, which was home to the 1924 Winter Olympics.

Keeping visitors happy

Facilities were also provided to let people take a dip even when the sea was too cold. The 1930s was the great age of the lidos, many of them boasting heated swimming pools. Like the cinemas of the day, lidos were designed with the smooth lines and curves of Art Deco, creating an atmosphere of glamour and luxury that reflected Hollywood films.

Holiday accommodation was something of a problem for those on a budget – hotels were too expensive and the cheaper alternatives could be grim. In 1935 a 36-year-old funfair operator named Billy Butlin, spotting a niche in the market, responded by setting up the first Butlins holiday camp. It opened the next year on former turnip fields on the outskirts of Skegness, offering an all-in deal that covered accommodation, free entertainment and three meals a day. It was a huge success, and other camps soon followed at Clacton in 1938 and Filey in 1945.

Touring country roads
The Citroën 2 CV, introduced in 1948, was the French equivalent of Germany's Volkswagen Beetle – a people's car that made motoring affordable for the not-so-well-off.

A passion for the countryside

Some holiday-makers had a taste for more vigorous leisure activities. The combination of increasingly sedentary jobs and growing amounts of leisure encouraged the spread of keep-fit movements. Within the cities women joined the League of Health and Beauty, which organised physical-education classes. In 1925 a health enthusiast set up the National Playing Fields Association to provide spaces for sport – an urgent need in a crowded country.

Other fitness-conscious city-dwellers preferred to spend as much of their spare time as possible getting away from it all. In a previous generation the novelist H G Wells had sung the praises of the bicycle as a means of escaping into the countryside, and many chose

THE BRITISH ABROAD

By 1930 around 1 million Britons were already taking holidays on the Continent each year, but they remained very much a minority of the population as a whole. The age of mass overseas tourism began with the first charter flights in 1950 and only really got under way in the 1960s. By 1970 the number of visitors to mainland Europe had risen to 6 million a year, reaching 10 million ten years later.

to follow his example. By 1938 the Cyclists' Touring Club and the National Cyclists' Union had more than 60,000 members between them.

Here again, a problem for those who wanted to venture farther afield was the lack of affordable overnight accommodation. To meet the growing need the Youth Hostel Association was created in 1930 specifically to provide beds for hikers, which it did for as little as a shilling (5p) a night. The response was overwhelming: by 1939 there were almost 400 hostels operating in mainland Britain and the movement had 50,000 members.

Up and away

Although the great outdoors attracted people from all walks of life, leisure time in Britain before 1945 remained very much divided along class lines. For the majority of those who could afford to take holidays, it meant conforming to the bidding of autocratic boarding-house landladies who expected their guests to be out of the house by 10 o'clock each morning and not to return before dinner time. Only the well-off could afford luxury hotels or overseas trips, usually to France or Switzerland.

The big change in holiday patterns came after the war, with the development of cheap air travel. From the 1950s on, package holidays began to reduce the cost of sun-and-fun vacations in southern climes. By the 1980s a generation used to foreign breaks were importing tastes and trends that would have been alien to most Britons in the 1930s. Patterns of recreation that had survived little changed since Victorian times suddenly found themselves under threat – and the seaside landlady was one of the revolution's victims.

Desert island dream
Created in 1950, the French holiday company Club Med offered customers increasingly far-flung locations. This tiny resort is in the Maldives.

A FUNDAMENTAL RIGHT

Article 24 of the Universal Declaration of Human Rights – adopted by the United Nations on 10 December, 1948 – affirmed the right of workers to have paid holidays.

RIGHT TO RAMBLE

Crowded into cities and suburbs during the week, outdoor enthusiasts in the interwar years displayed a passion for hiking at weekends. The Ramblers Association was created in 1935, partly to protect the rights of walkers and to press for open access to wilderness areas. Three years earlier matters had come to a head at Kinder Scout in the Peak District, where campaigners seeking enhanced rights of way staged a mass trespass. Six walkers were arrested, and five subsequently served prison sentences of between two and five months.

A breath of fresh air
Hikers in Normandy enjoy a view of Mont-St-Michel in 1937. The French firm Lafuna introduced lightweight backpacks in the 1930s.

Lift off for rotor-powered flight

Helicopters were developed over many decades, but only really took off as useful aircraft on the eve of the Second World War. Their exceptional manoeuvrability equipped them for delicate missions, whether civil or military.

The machine that lifted off at Hemelingen in Germany on 26 June, 1936, was an odd-looking contraption. It had a conventional canvas fuselage stretched over a wooden framework, but no wings. Instead, two metal outriggers each supported a large horizontal rotor. The twin rotors provided both the lift to raise the aircraft from the ground and the propulsion to drive it forward. They rotated in opposite directions, thereby countering the gyroscope effect that would otherwise cause the plane to pivot on its own axis. The craft, designed by Heinrich Focke, was in fact the world's first functional helicopter: it took off and landed vertically, and could maintain a stationary hover. It had a range of 200km (125 miles), a top speed of 120km/h (75mph) and could reach an altitude of 3,000m.

A long time coming

The Focke Fw 61 marked the final stage of a long process of development. A Frenchman, Gustave Ponton d'Amécourt, had coined the term 'helicopter' in 1831 from ancient Greek for 'spiral wing'. The principle involved had been spelled out even earlier, in the 18th century, by a Russian, Mikhail Lomonossov, who got the idea from windmill sails. D'Amécourt himself sought to use clockwork to turn rotors and in his wake many inventors experimented with rotary-wing machines, among them Frenchmen Paul Cornu and Louis Breguet and Denmark's Jacob Ellehammer. None succeeded because the engines at their disposal all lacked sufficient power.

Slow progress

A fresh generation took up the challenge at the end of the First World War, benefiting from the progress made in aircraft engines. In 1923 a Spanish engineer named Juan de la Cierva successfully flew an autogyro, a less complex machine than a helicopter that gets its forward thrust from a propeller, like an aeroplane, but receives vertical lift from a rotor turned by the upward flow of hot air. The problem with de la Cierva's machine was that in horizontal flight the craft's forward momentum gave the advancing blades greater airspeed than the retreating ones, generating unequal lift so the craft moved unevenly. De la Cierva's answer was a mechanism that automatically adjusted the pitch of the rotor blades, modifying the angle of incidence in accordance with their position. With this adjustment his autogyro successfully flew for short distances, and for a time it attracted the interest of British, French and American military engineers.

Among the pioneers of the late 1920s were Austrians Raoul Hafner and Bruno Nagler who designed and built a single-seater using a single rotor with a pair of fixed wings located

Early attempt
Paul Cornu and his prototype helicopter outside Lisieux in Normandy in 1907. The craft, mounted on bicycle wheels, moved forward on the day but probably did not fly.

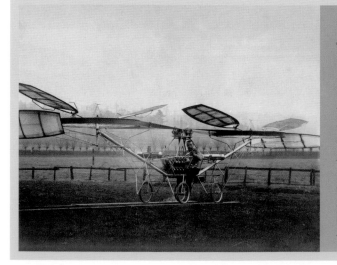

PAUL CORNU'S FLEA JUMP

The French claim that the first person to take off from the ground in a helicopter was a self-taught engineer from Normandy named Paul Cornu (1881-1944). He used his family's workshop, where they repaired bicycles and sewing machines, to build a motor-powered tricycle, an early motorcycle and then a small car. Excited by the dawning era of flight, he set out in about 1905 to build a helicopter employing a gasoline-powered engine to drive two horizontal rotors joined by a belt. According to Cornu's own account, he found himself lifted off the ground by the machine on 13 November, 1907, while trying to hold it down. Critics have since questioned the assertion, calculating that the device could never have become airborne, so the claim that it was the first helicopter flight remains unproven. Cornu was killed in his home during the Allied bombing of Lisieux that accompanied the D-Day landings.

FOCKE – AVIATION PIONEER

Heinrich Focke (1890-1979) was forced to interrupt his engineering studies by the outbreak of the First World War. With the restoration of peace he returned to his true vocation: designing aircraft. In partnership with Georg Wulf, an old friend, he started building flying machines in 1924. The two men established the Focke-Wulf aircraft company, which merged with Albatross in 1931. Thereafter Focke turned his attention to rotary-wing craft, producing de la Cierva autogyros under licence until he found himself too constrained by the limitations of the machine and decided to develop a genuine helicopter instead.

History intervened once more with the Nazi takeover of power in Germany in 1933. When the new rulers began to tighten their grip on the aircraft industry, seeking to prepare for war, Focke gave up the management of his company to concentrate on his old passion of aircraft design. He launched the Fw 61 helicopter in June 1936, but the new management at Focke-Wulf proved lukewarm about the project, so he set up a separate partnership with the test pilot Gerd Achgelis. In 1938 Focke-Achgelis produced the Fa 223 Drache, which saw only limited service in the war years but proved to be an inspiration for helicopter designers after 1945.

Meet the Fockes
The Foche-Achgelis Fa 223 Drache (above) had two main rotors installed on outriggers. It had a maximum speed of 182km/h (113mph). It was first produced in 1938 and about 40 machines were built between 1939 and 1945. Only two prototypes of Heinrich Focke's original Focke-Wulf Fw 61 were ever built, one of which – labelled D-EKRA – is shown here (left). It first flew in June 1936.

91

The first autogyro
Juan de la Cierva demonstrates his pioneering autogyro, equipped with both a rotor and a propeller, to British Air Ministry personnel at Farnborough in October 1925. The plane had made its first successful flight in Spain two years earlier.

in the downwash of air from the blade to keep the machine stable. The machine was not a great success, but Hafner made improvements to it after emigrating to Britain in 1932, where he developed further designs inspired by de la Cierva. The Hafner RII had flapping blades and was easier to control. (In 1940 Hafner was interned as an enemy alien; after the war he stayed in Britain, becoming a designer for the Aeroplane Company and later for Westland.)

At roughly the same time an Argentinian, Raoul Pateras Pescara de Castelluccio, was working on a craft powered entirely by rotors. He tried to resolve the technical problems of design by means of twin coaxial rotors. The problem of stability was eventually solved in

the 1930s, first by a little-known 'Laboratory Gyroplane' devised by two Frenchmen, Louis Breguet and René Dorand, and then by Heinrich Focke, whose Fw 61 was successfully tested in 1936. Three years later, in the USA, Russian-born Igor Sikorsky launched the VS-300, the first helicopter to have the now familiar large horizontal blade and small anti-torque tail rotor that stops the helicopter swinging around in the opposite direction to the main blade. The machines were about to go into production when the Second World War broke out. After the war Sikorsky helicopters were made under licence in Britain by Westland. Its Dragonfly (the Sikorsky S-51) made its maiden flight in 1948 and its debut with the RAF and Royal Navy in 1953.

Helicopters played only a limited role in the war. The German army tested the Flettner Fl 282 Kolibri, but no orders followed. The Focke-Achgelis Fa 223 Drache, successor to the Fw 61, did go into production but too late to have an impact. The US Sikorsky R-4 was put to use for rescue missions from 1944 on.

The turbine revolution

In the post-war years the industrialised nations threw themselves into developing new designs, but the fact remained that helicopters were

---

AUTOROTATION

Forward momentum makes a horizontal rotor turn freely of its own accord. This autorotation is enough to keep a craft airborne – the principle that permits autogyros to fly. The same effect ensures that if a helicopter's engine fails, the machine does not just fall straight out of the sky; by releasing the rotor the pilot can take control of what has become, in effect, an engineless autogyro, which in most cases should allow time to find a suitable place to land before the craft loses too much speed. Landing helicopters without power is part of pilot training.

HOW HELICOPTERS FLY

The helicopter's secret, and the source of its complexity, lies in the fact that each blade's angle of incidence automatically varies in the course of each rotation. These regular, cyclic alterations of pitch ensure stability of lift, but are modified every time the controls are employed to steer the craft. To climb, descend, accelerate or slow down, the helicopter pilot modifies the pitch of the rotor using a joystick-like device called the cyclic, which serves to tilt the rotor disc. There are also anti-torque rudder pedals – these control the pitch of the tail rotor blades, causing the helicopter to change direction.

VARIETIES OF ROTORCRAFT

Not all rotorcraft are helicopters. To qualify as a helicopter the craft must have a rotor powered by an engine, with the rotor alone responsible for keeping the craft in the air and moving forward. Autogyros do not qualify because their rotors are not powered – they are turned by the forward momentum of the craft. An autogyro also needs a conventional engine, typically driving a propeller, to get it off the ground. As a consequence, an autogyro can neither take off vertically nor hover.

There are also gyrodynes, which have motor-driven rotors like helicopters, but in their case the engines are turned off in flight: once the craft is airborne, it depends on conventional propellers to drive it forward and provide the speed necessary to keep the rotors turning. Gyrodynes are now mostly historical curiosities, largely due to their heavy fuel consumption.

To compensate for the generally low flying speed of helicopters – the maximum is about 250mph – American designers developed the Bell-Boeing V-22 Osprey tiltrotor, which has twin turboshaft engines driving two huge rotors. These are raised horizontally for take-off and landing, when they act like helicopter blades, but in flight the mountings are lowered 90° to align the blades vertically, so the plane flies as a conventional turboprop. The development programme of the V-22 experienced serious cost and time overruns, and its long-term future remains uncertain.

Staying airborne
Igor Sikorsky at the controls of the VS-300 on its maiden flight. In 1941 he established a new endurance record, remaining airborne for 1 hour, 36 minutes and 26 seconds.

woefully inefficient, eating up far more power than conventional aircraft of comparable size. Some machines could function on 1,500 or 2,000hp, but performance was disappointing. This situation was about to change.

In 1947 the French firm Turboméca invented turboshaft engines. Its Artouste model – first developed with 260hp and later improved to 380hp – had a very high power-to-weight ratio. Mounted in a Sikorsky S-59 helicopter, it broke speed and altitude records

Practical machine
A Sikorsky R-4 piloted by an instructor in 1945 as RAF officers look on (left). Of the 100 or so R-4s built for the US Air Force, 45 eventually found their way into British hands.

HELICOPTERS AT WAR

Helicopters in Vietnam
*Bell UH-1 Hueys (top) in action
in March 1968. The Boeing
CH-47 Chinook – seen here
(left) at the Battle of Khe Sanh –
was used for carrying troops,
supplies and wounded men.*

People who have seen the Francis Ford Coppola film *Apocalypse Now*, set in the Vietnam War, are unlikely to forget the helicopter gunships swooping down on a village. In difficult terrain, where airstrips were hard to come by, the US military turned to helicopters in Vietnam. Some 3,900 machines were mobilised, two-thirds of them Bell UH-1 Hueys (top). It did not take long for the Hueys to become symbols of the war. Originally intended for general support, they were soon equipped with cannons, machine-guns and antitank missiles. The UH-1's successor, the AH-1 Cobra, was a two-man attack helicopter, the first of its kind when introduced in 1967. It in turn was replaced in the 1980s by the Hughes Apache. Meanwhile, European-made machines had come into service, notably the Franco-German Tiger, but the numbers of these remain limited. Boeing CH-47 Chinooks (above) played an important part as troop transports in both the Korean and Vietnam conflicts and are still in use today in Afghanistan.

Adaptable warhorse

A US Marine Corps CH-53D Sea Stallion over desert terrain in Iraq's Al Anbar province in May 2007 (above). Designed by Sikorsky, the aircraft was originally intended for land-sea operations.

for the craft. Since then most helicopters have been equipped with turbine engines. The first to be mass produced was the American Bell UH-1 Iriquois – the 'Huey' – with a two-bladed rotor and a tail rotor. A modification of a 1954 design it was launched two years later; with its distinctive 'whomp-whomp' sound, it became symbolic of the war in Vietnam.

Today there are many different models of helicopter in varying shapes and sizes. In some ways their continued popularity is surprising, given that they are noisy, heavy, complex and costly machines, thirsty for fuel and with performance that hardly compares with other aircraft. The secret of their success, of course, is manoeuvrability. Taking off and landing vertically, they do not need an airstrip. They can fly in any direction and can also hover – abilities that make them uniquely suited for specialised tasks in hard-to-reach locations. The military quickly learned to appreciate these advantages, and civilians followed suit.

A multitude of uses

Helicopters first became essential military hardware in the Korean War (1950–3), when they were used to move troops and evacuate the wounded. The French put them to similar

use in Indochina (1946–54) and Algeria (1954–62), fitting some with machine-guns and rockets. The US military made heavy use of helicopter gunships in the Vietnam War (1955–75). Helicopters are regularly used in military situations where speed is not the prime consideration, for instance in attacking ground targets, submarines or tanks.

In the civil sphere, helicopters play a role in mountain and sea rescues, medical emergencies and fire-fighting. They are used for short-range travel and also as flying cranes that can carry heavy loads to inaccessible places. In the 1960s aviation companies toyed with the idea of developing fleets of extra-large helicopters to provide passenger services, but this was ruled out due to the heavy fuel consumption and lengthy flight times involved.

Eurocopter EC 145

In service since 2002, the EC 145 has gradually replaced the Alouette III.

Artificial hearts 1937

Heart maker
Dr Robert Jarvik shows off the Jarvik 7 artificial heart in 1981. The organ, which was made largely of polyurethane, had completely smooth artificial ventricles to reduce the risk of blood clots.

In Moscow in 1937, when the Stalinist terror was at its height, Vladimir Demikhov, a 21-year-old biology student, found time to develop the first artificial heart. He tested it on a dog, and even though the pump was too large to fit inside the animal's thorax, the dog nonetheless survived for five and a half hours. Demikhov continued with his experiments on dogs, subsequently carrying out the first heart and lung transplants.

Two decades passed before research on artificial hearts resumed, this time in the hands of a Dutch-born US citizen named Willem Kolff. Kolff already had pioneering work on kidney dialysis to his credit when, in 1957, he implanted a plastic heart with an air pump into a dog; the animal survived for an hour and a half with the plastic heart. Before experiments on humans could get under way, a reliable electric motor had to be developed, along with some suitable external control system. It was also vital to determine the most suitable materials to use, principally to avoid the risk of blood clots.

In all some 250 people collaborated in this ambitious project, which finally bore fruit in 1969, two years after Christiaan Barnard had carried out the first human heart transplant using a real heart. That year, a surgeon named Denton Cooley implanted an artificial heart designed by Domingo Liotta into a patient waiting for a transplant. The artificial heart operated for 64 hours before it was removed for the transplant operation to take place; sadly, although the artificial heart had performed well, the transplant with a real heart proved unsuccessful.

Another major step forward came in 1982, when a Jarvik 7 device – named after its designer, a colleague of Kolff's – was implanted in a patient named Barney Clark, a retired dentist. Clark survived for 112 days with the aid of the artificial organ, which was attached to a cumbersome control apparatus weighing 40kg (90lb).

Today doctors are still looking for ways of improving artificial hearts using miniature electronics, whether to serve as long-term replacement organs or as stop-gaps for patients awaiting permanent transplants.

PROMISING PROGRESS

Recent progress in the fields of miniaturisation and electronics have raised hopes that it might soon be possible to develop an implantable organ employing living cells derived from stem-cell culture. French surgeon Alain Carpentier has developed a biosynthetic heart; a compact prototype should be ready for clinical trials in 2011. Researchers in the USA, Japan and South Korea are racing to produce their own versions.

State of the heart
Alan Carpentier demonstrates his prototype artificial heart, made of biocompatible materials and incorporating complex electronics.

Inflatable boats 1937

Lifesaving craft
*Inflatables are
popular with
lifeguards, being
fast and easy to
manoeuvre even
in rough seas.*

The French Zodiac company was set up in 1896 to build airships. In the 1930s one of its employees, Pierre Debroutelle, was looking for new products to develop. In 1934 he came up with an idea for a pneumatic rubber boat consisting of two blown-up tubes linked by a rubberised canvas hull. Three years later he added a wooden instrument panel at the stern. The French navy subsequently adopted the craft to transport bombs and torpedoes. Unlike the wood or metal dinghies used up to that time, the Zodiac's U-shaped inflatable tube had the advantage of not damaging seaplanes during loading or unloading.

A popular leisure craft

During the war, Debroutelle worked secretly to improve his prototype, using scrap material from airships. He added a motor in 1940, carrying out the first trials on the River Seine. After the war the boats enjoyed extraordinary success. Zodiacs accompanied the French marine explorer Jacques Cousteau on his expeditions. Today they are made of synthetic materials covered in neoprene.

In 1969 a British Admiral named Desmond Hoare patented the first inflatable boat to have a rigid plywood hull, which had first been made two years previously by Tony and Edward Lee-Elliott. The inflatable collar, which allows the boat to remain buoyant even when much water is shipped aboard, has made it invaluable in sea rescue.

SEA SURVIVOR

In 1952 Alain Bombard, a 27-year-old hospital doctor from the French port of Boulogne, set out to prove that it was possible to survive alone at sea for several weeks. To do so he aimed to cross the Atlantic single-handed in a rubber inflatable, equipped with a sail but without water or provisions. Bombard – seen here with his friend Jack Palmer, who accompanied him as far as Tangier – made the trip in a Zodiac called *L'Hérétique*, surviving on what he could catch by fishing and by harvesting surface plankton; he drank rainwater and liquid pressed out of the fish he caught. By the time he arrived in Barbados after 64 days at sea, he had lost 25kg (55lb) in weight, as well as his fingernails and toenails, and he was suffering from serious eye and skin disorders. He published an account of his voyage the following year.

Petrochemicals compete with latex

To free themselves from dependency on countries that produced natural rubber, first Germany and then the USA sought to develop a synthetic substitute. The progress of polymer chemistry and petrochemicals gave them the means, and synthetic rubbers were soon being put to industrial use.

Rubber plane
Twelve prototypes of the Goodyear Inflatoplane (right) were produced for the US army between 1955 and 1972. This revolutionary craft could be packed in a container and dropped by parachute.

Making tracks
A Buna S synthetic rubber tyre is mounted for display in Berlin in 1937, at an exhibition designed to promote German industry and technology.

At the start of the 20th century, natural rubber was obtained from latex extracted from plants. Thanks to the vulcanisation process, which involved adding sulphur to the gum and then heating it to make it elastic and heat-resistant, rubber became an indispensable ingredient in many industrial products, particularly tyres. It was so successful, in fact, that demand outstripped production, despite the development of large plantations of rubber trees in the tropics, and prices soared.

An artificial substitute

In Germany Fritz Hofmann, a chemist for Bayer, laboured to find a synthetic substitute. It had been known since 1905 that natural rubber is a polymer formed by long chains of molecules of isoprene, a common organic compound that could not be replicated in laboratories at the time. From 1906, Hofmann began testing compounds extracted from coal tar that were chemically close to isoprene, heating them to different temperatures in tin cans in an attempt to polymerise them. In 1909 he took out a patent on a synthetic rubber produced from methyl isoprene.

The following year Carl Duisberg, Hofmann's employer, drove 4,000km (2,500 miles) in a car fitted with tyres of the new material without suffering a single puncture. But there were problems with the product: it was difficult and costly to make, and for want of natural stabilising agents it was subject to wear and tear, easily becoming oxidised. Production came to a halt in 1913.

The path to success

Even so, the chemistry of elastic polymers, the so-called elastomers, was making progress. Researchers sought to stabilise them with additives and tried out new processes to

---

AN ESSENTIAL WAR MATERIAL

In the years between the wars, the growing number of rubber plantations in the colonies slowed the development of synthetic rubber by lowering the price of the natural product. That situation changed radically with the shortages caused by the Second World War, when Germany found itself under blockade and the USA was cut off from its Far Eastern suppliers by the Japanese. The two nations were soon producing industrial amounts of synthetic alternatives.

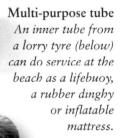
Multi-purpose tube
An inner tube from a lorry tyre (below) can do service at the beach as a lifebuoy, a rubber dinghy or inflatable mattress.

transform various hydrocarbons into rubber. In the late 1920s Germany's Walter Bock hit upon the idea of using sodium as the catalyst to link butadiene with styrene. The result was styrene–butadiene rubber (SBR), which was patented in 1929 and subsequently marketed under the trade name Buna S – *Bu* for butadiene, *na* for natrium (the German word for sodium) and *S* for styrene. Proving much more durable than methyl isoprene rubber, SBR was first used to make car tyres in 1935. By that time Eduard Tschunkur and two other chemists employed by IG Farben were developing Buna N, or nitrile rubber, which had improved resistance to oil.

The US contribution

Meanwhile, there was strong competition from the USA. Two employees of the DuPont chemical concern, J A Nieuwland and Wallace Carothers (who was known as the inventor of nylon), had polymerised chloroprene to produce neoprene in 1930. The new substance was chemically stable and resistant to heat. Then in 1937 Robert Thomas and William Parks, while working for Standard Oil, developed butyl rubber by combining isobutylene with butadiene or isoprene. While not especially shock-resistant, butyl rubber proved almost completely airtight, making it very suitable as a sealant. It went into industrial production in 1941.

Various synthetic rubbers have been produced since, but none has proved as internationally popular as SBR, which continues to make up more than half of world production of the artificial product, largely for use in tyres. Neoprene (used in glues, rubber gloves and waterproof goods) and butyl rubber (mostly employed in tubeless tyres and inner tubes) are also well established.

SELF-REPAIRING RUBBER

Developed in France by researchers at the National Centre of Scientific Research (CNRS) and produced by the Arkema group since 2009, self-repairing rubber mends tears simply by coming into contact with them. A product of so-called supramolecular chemistry, it works through the action of tiny molecules derived from fatty acids that naturally bond together.

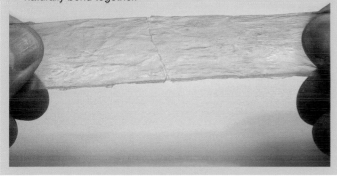

Father of the digital age

Alan Turing's name is indelibly associated with the idea of a 'universal' machine – a programmable calculator that foresaw today's computers. His other great achievement lay in deciphering the Enigma code used by German forces in the Second World War.

In 1936 Alan Turing, a fellow of Kings College Cambridge, outlined the principle of the Turing machine, a universal computer able to perform calculations on data fed in on perforated tape. He was the first person to discuss recursively calculable functions – that is, mathematical problem-solvers that carry out complex tasks by performing lengthy sequences of elementary operations, also known as algorithms.

The idea of a perfect computer

To perform such operations, the Turing machine presupposed a memory capable of retaining the entire sequence of algorithms pertaining to the problem it was set to solve, as well as entry and exit mechanisms to feed it raw data and to deliver the results. Its range was effectively limitless so long as algorithms could be found to express the functions to be calculated. The device, which worked through a succession of separate 'states' or intermediate steps to arrive at its conclusions, effectively foreshadowed the arrival of the binary system, employing the symbols 0 and 1, used by

modern computers. But unlike real-life computers, which are naturally limited by the resources available to them, Turing's theoretical machine had unlimited input and so was truly universal.

Codes to decipher

In 1939 Turing was seconded to Bletchley Park in Buckinghamshire, the main decryption centre dedicated to breaking the ciphers used by enemy forces in the Second World War. Foremost among these were the codes produced by the German Enigma machines, which employed a sequence generated by rotors to encode messages automatically. The word MESSAGE itself, for

Enigma machine
The encryption and decryption devices adopted by the German military had been available commercially since the early 1920s. They featured from three to eight separate rotors (inset, left).

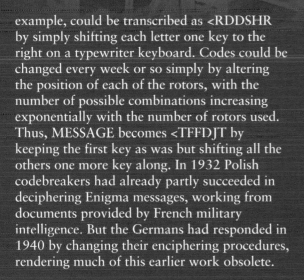

example, could be transcribed as <RDDSHR by simply shifting each letter one key to the right on a typewriter keyboard. Codes could be changed every week or so simply by altering the position of each of the rotors, with the number of possible combinations increasing exponentially with the number of rotors used. Thus, MESSAGE becomes <TFFDJT by keeping the first key as was but shifting all the others one more key along. In 1932 Polish codebreakers had already partly succeeded in deciphering Enigma messages, working from documents provided by French military intelligence. But the Germans had responded in 1940 by changing their enciphering procedures, rendering much of this earlier work obsolete.

The Enigma mission

Turing's method consisted of focusing on those elements in any given message that were easiest to decipher. Four consecutive characters, for example, might well stand for a year, say 1941; two repeated letters might also provide a clue, because only a limited number of words display this feature. By working along these lines, it could become possible to work out the position of the encoding machine's rotors and to devise a way of deciphering its output even if the configuration of the rotors was subsequently changed. In theory the number of possible combinations was astronomical – the figure has been calculated at over 150 trillion   but in practice some juxtapositions of letters could be

ruled out from the start, considerably reducing the work involved. The task then became a matter of calculating probabilities.

In 1940 Turing and his colleagues created a counter-Enigma machine with 624 electronic connections that was capable of deciphering German messages in two or three hours, working at a rate of 20 operations a second. Thanks to this famous 'Turing bombe', several copies of which were made, as well as to information provided by the capture of Enigma rotors from a German submarine, the code was finally broken, providing the Allies with a decisive military advantage.

With the coming of the digital age, deciphering techniques have evolved from relying on substituted combinations of letters to sequences of bits (*binary digits* with a value of 0 or 1) encrypted using techniques classified by experts as either 'symmetric' or 'asymmetric'.

Codebreaker
Operators service a bombe, as the cryptanalytic machines designed by Turing were known. There were two of the code-breaking machines at Bletchley Park, where a team of 7,000 people, hand-picked for their problem-solving abilities, worked to decipher German military messages.

Instantaneous copies on demand

The coming of the industrial age vastly increased the amount of paperwork in circulation, from order forms and delivery slips to contracts and patents. And as trade grew, the need for copies of documents also grew. Responding to this demand, the photocopier finally did away with the chore of producing duplicates by hand.

The world's first photocopy

In 1938 Chester Carlson (above) and his associate Otto Kornei made the world's first photocopy – of a microscope slide (above right) – using a zinc plate coated with sulphur. Carlson had taken out a patent on the process the previous year.

The duplication techniques available in the early decades of the 20th century – for instance lithography, offset printing and the rotary press – were ill suited to reproducing small numbers of copies. Meanwhile, the printing methods that were suited to limited numbers of copies, such as mimeographs or Roneo duplicators, required a stencil of the original document to be prepared by hand.

In 1907 George Beidler invented a machine to photocopy documents and develop them on paper, founding the Rectigraph Company in Oklahoma City to market his device. But the apparatus was cumbersome and difficult to use, resembling more a photographic laboratory than anything else. To take a copy a technician had to be summoned, returning with the document days later.

A taste for invention

Chester Carlson was working as a clerk in a US patent bureau when he put his mind to finding a better solution. As a student of physics he found it surprising that no-one had found a commercial use for static electricity.

Exploring the idea, he discovered that the pull of static is sufficiently powerful to retain ink in powdered form, and also that photons of light can cancel out the charge. It followed that if the image of a document could be projected onto a drum charged with static electricity, only the written parts of the document would remain negatively charged and so pick up ink. Carlson patented the process, which he called electrophotography, in 1937. With the assistance of a young Austrian physicist named Otto Kornei, who helped him to surmount the various technical problems involved, he produced the first photocopy on 22 October, 1938. Carlson then offered his invention to some 20 different companies, all of whom turned it down as non-viable.

Success at last

Eventually the Haloid Corporation of New York agreed to put some money behind the project. Buying up the patent in 1947, it renamed the process xerography, from the Greek for 'dry writing'. Later this small manufacturer of photographic paper would change its name to Xerox.

The first models that the firm produced needed special paper that wore badly and were not particularly successful. It was only with the introduction of the Xerox 914 machine, which printed on ordinary paper, that photocopying really took off. Gradually photocopiers became standard equipment in offices and schools, replacing earlier, so-called 'wet' copying methods. The Xerox photocopiers, for example, killed off the old spirit duplicators, which used solvents largely composed of alcohol and whose copies sometimes came out crumpled from the rollers. The new machines delivered black-and-white copies instantaneously to a standard quality. Copies were also relatively cheap, so long as the numbers were limited. The Japanese firm Canon introduced colour copying in 1973.

Over time photocopiers took on new roles, such as stapling and binding documents or acting as scanners, and were modified to hold different sizes of paper in separate drawers. Connected to computer networks, they came to serve as office printers and fax machines, sending documents by electronic mail.

The reverse of the coin was that photocopiers vastly increased paper consumption and made the illegal copying of copyrighted material much easier. By the late 20th century the quality of some copies was so good that central banks were forced to incorporate holograms and other supposedly 'uncopiable' features into banknote design to prevent counterfeiting.

Early Xerox
The first Xerox photocopiers went on the market in the 1950s. This model (left), the Xerox 1385, was introduced in about 1960. By the following decade Xerox had grown into a multinational concern with 95 per cent of the world photocopier market.

HOW A PHOTOCOPIER WORKS

The xerographic process consists of charging a drum with static electricity, then projecting onto it an image of the document to be copied. The flow of light photons cancels out the electrostatic charges on all parts of the drum except those corresponding to the black or written sections of the document. The ink, in the form of dry powder, fixes itself only on those parts of the drum that remain negatively charged. When a blank sheet of paper is pressed against the drum, the ink is pressed onto it to form an exact copy to the original document. The ink is then 'baked' into place by heat and pressure rollers so the copy will not fade.

Maintenance and repair
With the coming of the internet, linked-up copiers could be set up and adjusted from a distance via the Web.

The Volkswagen Beetle 1938

With 21.5 million cars sold since it was first introduced, the Volkswagen Beetle beat all records in automobile production. No other model has sold so many cars in such a recognisably unchanged design.

The Beetle was born of Hitler's desire for a 'people's car' – the literal meaning of the name Volkswagen. The design brief specified that it had to be solidly built, large enough to hold a couple with three children, have a cruising speed of up to 100km/h (60mph) and use less than 7 litres of petrol per 100km – and all this for a maximum of 1,000 marks, which was just over 30 weeks' salary for the average German worker at the time. Ferdinand Porsche took up the challenge and devised various prototypes. The KdF Wagen – short for *Kraft durch Freude*, 'Strength through Joy' – was launched in 1938. Few civilian cars were made in the war years, but the chassis was put to use to make vehicles for the German army.

The Beetle came into its own after 1945, when the British occupation authorities in Germany ordered 20,000 units. Ten years later more than a million had been sold, an unprecedented figure for the European car market. Exported to almost 150 countries, the Beetle evolved through many phases with little change to its outward appearance. It got a 30hp engine and a single-piece rear window in 1953, a 34hp motor in 1961, a new design with slightly more window space in 1964, a fresh chassis with a 40hp engine in 1966, increased again to 50hp in 1970. In 1972 it took the world record for volume production from the Model T Ford, with over 15 million vehicles sold. Six years later, Beetles were being produced in Brazil and Mexico and the *sedan clasico*, as the model was known, was the cheapest car on the Mexican market. A New Beetle was launched in the USA in 1998. Production of the record-breaking model ceased in July 2003.

FPM 422C

Iconic design
This model (above) was produced in 1965, almost three decades after the rounded, beetle-shaped lines had first been drawn up.

AN UNLIKELY FILM STAR

Over time the 'people's car' acquired an unlikely cult reputation. Dubbed the Beetle for its quirky design, it inspired an affection only felt for a handful of other equally eccentric vehicles, the Mini and the Citroën 2CV among them. Beetle-mania even reached Hollywood in the shape of Herbie, the hero of Disney's *Love Bug*, made in 1968. Four sequel films followed, plus a 1997 made-for-TV sequel to the original, also known as *The Love Bug*.

The magic million
On 5 August, 1955, workers at Volkswagen's Wolfsburg production plant joined more than 1,000 Beetle enthusiasts to celebrate the moment when the millionth car rolled off the production line (left). Coming a decade after the end of World War II, the event was seen as a symbol of Germany's post-war economic miracle.

Superman 1938

Superhero's progress *Superman made his first appearance in Action Comics (below) in 1938. By the following year he had a comic book out in his own name (bottom).*

In 1938 an American comic-strip monthly, *Action Comics*, had an athletic new character on its cover dressed in a blue suit with red cape and boots. The large 'S' emblazoned on his chest stood for his name: Superman. The new hero could fly through the air as fast as a speeding bullet and had the strength to lift a train from the tracks just before it was derailed – all because, it was explained, the heavier gravity on the planet he came from had boosted his musculature. He had x-ray vision that could penetrate walls and his body was protected by a magnetic field.

Nietzsche comes to the comics

Superman was the creation of Jerry Siegel and Joe Schuster. He was born Kal-El on the planet Krypton, but on Earth he took the persona of Clark Kent, a journalist on the *Daily Planet* who had been raised by a Kansas farmer and his wife. Driven to redress the wrongs that threatened the world, he soared through the air to stop planes dropping out of the skies.

Siegel and Schuster's character was an immediate success and Superman became an iconic figure of popular culture, spawning a whole generation of new superhero imitators. Before long, the list would include Batman, Spiderman and the Incredible Hulk, as well as female counterparts such as Wonder Woman. Meanwhile, Superman himself found a new incarnation in the cinema, most famously personified by Christopher Reeve.

People's need for legendary heroes goes back a long way, stretching at least as far as Ulysses who killed, the one-eyed giant Polyphemus of Greek myth, and St George who dispatched the dragon. Superman, the archetypal caped crusader, brought a modern twist to this tradition through the canny appliance of 20th-century science.

COMIC STRIPS

Comic books trace their history back to albums drawn by the Swiss caricaturist Rodolphe Töpffer, a friend of the German poet Goethe, in the 1830s. Newspapers began running comic strips in the last years of the 19th century; the hugely popular Katzenjammer Kids first appeared in a New York paper in 1897. The British press was relatively slow to adopt the format, with well-known characters like Andy Capp and Flook only rising to prominence in the 1950s. Otherwise UK comic strips tended to be targeted at the young, in publications like the *Dandy* (home of Desperate Dan and Beryl the Peril) and the *Beano*, with Dennis the Menace and the Bash Street Kids.

A primeval survivor

The coelacanth has excited debate among biologists ever since its chance discovery. Is it a missing link in the evolutionary chain that saw sea creatures migrate onto land hundreds of millions of years ago?

In December 1938 fishermen in the estuary of South Africa's Chalumna River caught a strange fish. It had a huge mouth, blue scales and flippers resembling embryonic limbs. It looked so odd that when the trawler docked in the port of East London, northeast of Cape Town, the captain called Marjory Latimer, curator of the town's tiny natural history museum, to take a look. She knew they had something unusual, possibly prehistoric, and so contacted amateur ichthyologist James Smith, a professor at Rhodes University in Grahamstown, for a second opinion. When he got to see the fish the following February, Smith had no hesitation in declaring it a coelacanth, a species thought to have been extinct for 70 million years.

Ancestor of land-dwellers

Smith believed that the lobe-finned coelacanth was a missing link between fish and tetrapods – four-limbed, land-living vertebrates that include amphibians, reptiles and mammals. In his view it was the last living representative of the crossopterygians, fish that first appeared more than 400 million years ago and evolved to form the first land-dwelling tetrapods. In 1952 a second coelacanth was caught off the Comoro Islands, and since that time more than 200 others have come to light, including a variant species found in 1997 in Indonesia. The coelacanth's vertebrate characteristics – such as the limb-like flippers, vestigial lungs and a bony skeleton – fed scientific debate. Yet scientific opinion today is that the creature is not the direct ancestor of mammals that Smith believed it to be, but rather at best a distant uncle of the famous missing link.

FILE CARD FOR A LIVING FOSSIL

- **Latin name:** *Latimeria chalumnae*
- **Popular names:** Coelacanth, *gombessa* (in the Comoros), *ikan malam* (in Indonesia)
- **Length:** Up to 3m (10ft) long
- **Weight:** 60–95kg (130–210lb)
- **Distribution:** Principally in and around the Comoros archipelago.
- **Behaviour:** Observed in 1987 in its natural habitat, it was found to live in caves between 100 and 400m (300–1,300ft) down. It fed at night on fish detected with the aid of an electroreceptive rostral organ in its skull.
- **Conservation status:** Endangered, with an estimated total population of under 500.

Under threat
Scientists are pressing for the coelacanth species to be better protected. About 10 coelacanths are accidentally caught in fishing nets each year in the Comoro Islands, a heavy toll for a population estimated at less than 500 individuals.

NYLON – 1938

The mother of synthetic fibres

Nylon was born of the chemical industry's efforts to develop artificial silk in the years between the wars. One of the first fruits of the polymerisation process, it proved an immediate success in the form of nylon stockings and it later turned out to have many other uses in varied applications, from flooring to heavy industry.

In 1928 Charles Stine, director of DuPont's chemical department, employed a young Harvard professor named Wallace Carothers, whose dynamism, team spirit and passion for research marked him out as someone of enormous potential. Stine obviously considered his new recruit a highly promising experimentalist, but he could hardly have guessed that he would be responsible for some 50 separate patents, nylon among them, that would turn DuPont into a giant of the global chemical industry in less than a decade.

Carothers was barely settled in the job before he applied his attention to the challenge of devising an artificial silk. The natural product was expensive, depending as it did on the long labours of silk worms, which were regularly decimated by epidemic diseases like phylloxera. At the time the only available synthetic fibre was rayon, manufactured from cellulose since the 1890s. DuPont had acquired the rights in the 1920s, producing the material in its factories. But cellulose itself was derived from wood, a natural substance, so rayon was not really a fully synthetic material.

Promising polymers

At Harvard, Carothers had been working on polymers – extra-large molecules made up of chains of smaller ones. Carothers saw these as the key to artificial silk. In April 1930 one of his assistants stumbled upon a solid polymer that could be drawn out to a great length without snapping. Unfortunately its melting point was low and it dissolved in water. Any fabric made of the stuff would not stand up to washing and risked melting when ironed. Even so, the discovery confirmed to Carothers that he was on the right track.

At the time his laboratory was also synthesising neoprene, used to make one of the early man-made rubbers. The economic crisis that had begun in 1929 now played a part. DuPont responded to the Depression by betting on artificial silk to restore its fortunes, and Carothers found

Beginnings of an industry
The very first sample of nylon (left), as developed by Wallace Carothers and his team at DuPont Chemicals in 1935. Industrial manufacture got under way in 1938. After the polymerisation process, the resultant liquid was extruded in fibres that were wound onto bobbins (above).

Indispensable accessory

There were riots in the street when nylons first became available. Everyone wanted them – this queue of enthusiastic shoppers (far left) is outside a store in Washington in the 1940s. When first introduced nylon stockings replaced rayon, a coarser material derived from wood pulp. Nylons owed their success to their smoothness, transparency and legendary robustness.

THE NYLON STOCKING SAGA

Transparent, light and exceptionally hard-wearing, nylons were an immediate success when they came on the market in 1940, quickly replacing the heavy lisle cotton and opaque rayon or viscose stockings that had been worn over the previous decade. Having won over the American market, they crossed the Atlantic toward the end of the war and soon were just as popular in Britain. Fashion at the time demanded a straight seam all the way down the back of the calf. By the time the miniskirt arrived in the early 1960s, the seamless stocking, made possible by the invention of the circular knitting machine, was already well established, and the invention of Lycra in 1959 had made it possible to create snug, stretchy 'one-size' hosiery. Thus decency, to avoid stocking tops being on display, caused suspender belts to be banished in favour of more practical tights, which had long been worn by dancers.

Under the microscope *Coloured nylon fibres like these (left) are relatively little affected by dampness or mould.*

himself under increasing pressure to deliver. Even though he was taking out patents on a growing number of polymers, he still had not managed to find a silk substitute. Then, on 24 May, 1934, one of his assistants finally identified a polymer whose melting point was not too low. Success was at hand. Some adjustments had to be made, but by February 1935 Carothers was able to present the firm's executive board with a polymer, prepared from adipic acid and hexamethylene, that was perfectly suited for weaving, being both hard-wearing and stretchable. The result was nylon, known at the time as Fibre 6-6 and patented in 1938.

Sadly, Wallace Carothers did not live to see the product's success. Long plagued by depression, on 27 April, 1937, he checked into a hotel room in Philadelphia and committed suicide by drinking lemon juice that he had laced with potassium cyanide.

From strength to strength

Once the problems of manufacturing had been ironed out, two factories were set up to meet demand for the new material, one in 1938 in Delaware, the other three years later in Virginia. In May 1940 the first nylon stockings went on sale, quickly becoming a star performer for DuPont. And there soon turned out to be plenty of other applications for the material, from toothbrushes and fishing lines to men's ties and football jerseys. In one year the company sold $25 million worth of nylon products. In 1941 a new market opened up for parachute fabric to supply airborne troops; over the course of the war, 3.8 million units would be manufactured.

After 1945 nylon came to be used principally in three sectors: for clothing, floor coverings and in industry. By the late 1940s the first nylon rugs and fitted carpets were winning a reputation for durability. In industry, nylon fibres were used to make tyres more hard-wearing. When melted and moulded, the material formed smooth pieces that were resistant to rubbing and found many uses in packaging for the food industry.

Over the years nylon has remained a standout performer for DuPont, even though competition was later provided by a number of nylon derivatives, such as Kevlar. Developed in 1971, Kevlar had such exceptional toughness it was suitable for making motorcycle helmets, bulletproof vests and similar products. Today many other polyesters, such as Reemay, Typar and Tyvek, have found a use in daily life or in industry, making nylon the matriarch of a whole family of synthetic fibres.

A new use in the skies
Supported by nylon canopies, parachutists of the 82nd Airborne Division float down to earth at the Fort Bragg military base in North Carolina sometime in the 1940s.

NYLON – ORIGIN OF A NAME

Over the years many bizarre stories have circulated about the origins of the word 'nylon'. One of the strangest, dating from the war years in the US, made it an acronym for 'Now You Lose Out, Nippon' – a notion that achieved such wide circulation at the time that DuPont had to issue an official denial. Another, equally mistaken theory held that it derived from 'New York–London', the two cities where the product was first launched. More romantic was the idea that it was made up of the first letters of the Christian names of its inventors' wives. In fact, the term first surfaced as 'no run', a name chosen by committee from 400 different suggestions. This was later adapted to 'nuron', then finally to 'nylon' so that the pronunciation would be the same on both sides of the Atlantic.

Pressurised cabins 1938

In 1938 cabin pressurisation was successfully introduced on the Boeing 307 Stratoliner, a four-engined airliner developed from the B17 Flying Fortress bomber. There were substantial advantages to be had from flying at high altitudes, as turbulence was reduced and fuel consumption was lower. High-altitude flying was only possible if conditions inside the cabin could be made comfortable and safe.

The higher that planes flew, the lower the air pressure and temperature inside the plane, eventually threatening to cause fatal embolisms.

An aviation revolution

Pressurisation involved using a compressor powered by the plane's engine to raise the pressure of air taken from outside the aircraft to an acceptable level – say, that found at an altitude of 2,000m (6,500ft). The pressurised air also had to be heated, for air temperatures at high altitude are extremely low, falling to –36°C (–33°F) at 6,000m (20,000ft). To maintain the pressure, the cabin had to be completely airtight. From 1936 the DuPont company started manufacturing tape made of neoprene (invented a few years earlier by J A Nieuwland) that effectively sealed the joins in the Lockheed XC-35 experimental aircraft used for the first pressurisation test flights.

Pressurised cabins were a decisive step forward in the onward march of aviation. Their spread was helped by the arrival of jet engines, whose turboreactors produced huge amounts of compressed air at very high temperatures, making both pressurisation and air conditioning much easier to achieve.

Testing pressure
The ten passengers on this Northwest Airlines flight (left) were taking part in a trial in 1939 to test the new cabin pressurisation. They were able to take off their oxygen masks without ill effect even after the plane had reached an altitude of 6,000m (20,000ft).

CABIN CREW

In the early 1930s some airlines flying the Paris–London route and to and from India added stewards to their crew to serve meals to the passengers. At the time planes flew slowly and at low altitudes where they were subject to much turbulence. The result was that those flying on them often got airsick, leading Boeing Air Transport (today's United Airlines) to start employing nurses. The first on the job was Ellen Church (right), who flew on the Chicago–San Francisco route. Most American airlines duly followed suit, and in 1935 TWA started calling these staff members 'air hostesses'. Today stewards of both sexes make up aircraft cabin crew, whose principal role is to ensure the safety and comfort of the passengers.

Instant coffee 1938

In 1938 Nestlé surprised coffee-drinkers with Nescafé, the first soluble instant coffee. At the start of the decade Brazil's coffee-marketing board, ever eager to sell more produce, had encouraged the Swiss firm in the development of cubes that would provide a cup of coffee when dissolved in hot water. After eight years of research the company decided to take a different tack: after the coffee beans had been roasted and ground, they were placed in huge vats and dehydrated to preserve them in powdered form. The product still had deficiencies at the time of its launch, relying on sugar additives to preserve its aroma, and despite the enthusiastic endorsement of American GIs during the Second World War, it failed to win over the public at large. It was only after 1952, when Nestlé improved its method of preconcentrating the aroma, that the sugar content could be removed and Nescafé could claim to be 100 per cent pure coffee.

Global favourite
A woman prepares instant coffee in a log hut in the Amazon rain forest.

Breathalysers
1938

Dr Rolla N Harger was a professor of biochemistry and toxicology in the medical school at Indiana University when, in 1938, he took out a patent on a machine he called the Drunk-o-meter, designed to measure the amount of alcohol present in an individual's blood. The device contained a small amount of orange-coloured potassium dichromate which turned green when it interacted with ethyl alcohol through a so-called 'redox reaction'. The alcohol in a person's breath is roughly 2,000 times less concentrated than alcohol in the blood, so from this it was possible to work out roughly how much a person had drunk by getting them to breathe into the balloon contained within the machine.

Harger designed his device as a first step in identifying motorists suspected of driving while drunk. If the kit duly turned green, other, more accurate tests, such as a blood test, could then be applied. A law was passed and Indiana quickly saw a reduction in the number of alcohol-related accidents. Sixteen years later

Testing, testing
Dr Harger testing his Drunk-o-meter in 1937 (left). The yellow crystals inside the device (below) change colour when someone blows into it. If the alcohol content is more than 0.5g/l, they turn green.

another Indiana University professor, Robert Borkenstein, introduced an improved device, more compact and reliable, that was duly called the breathalyser.

A SELF-ADMINISTERED TEST

Self-administered kits are now available that measure breath alcohol with the aid of an electric current. The person being tested blows into the tube through a disposable nozzle. Traffic light colours then indicate the amount of alcohol on the breath: green means all is well, orange 'Watch out', while red indicates over the limit.

Calling on chemistry to kill insects

When he first discovered the insecticidal properties of DDT, Paul Hermann Müller thought that he had come up with an infallible miracle weapon against typhus, malaria and crop-destroyers. But the pesticide's harmful effects on the environment soon became apparent.

Fighting vermin
US soldiers spray DDT on their clothes in 1944 (right). Vermin-borne disease was a persistent threat to the troops.

Spraying DDT
In 1944 an epidemic of typhus swept through Naples, exacerbated by wartime overcrowding. Liberal use of DDT brought the outbreak under control.

After graduating from the University of Basle in 1925, Paul Hermann Müller took a job with the Swiss pharmaceutical firm, Geigy. At first he conducted research on dyes, but ten years later he turned his attention to insecticides – poisons designed to kill insects that carry diseases and ravage crops, threatening agricultural productivity. The insecticides that had been developed in the 19th century – some of them employing arsenic, others derived from copper or kerosene – were not very effective, opening a market for new products.

A powerful weapon

After testing hundreds of compounds, Müller decided to concentrate on organochlorides, in particular dichloro-diphenyl-trichloroethane, better known today as DDT. The substance had first been synthesised by Othmar Zeidler in Germany in 1874, but had not aroused much interest until Müller discovered its insecticidal properties in 1939. It then became an instant success. From 1943, the US army used it to control epidemics; houses in Casablanca were sprayed with the substance to kill malaria-carrying mosquitoes, and in Naples it was used against lice that were spreading typhus. It was also widely used in high doses for crop-spraying.

Discovering the downside

In 1948 Müller was awarded the Nobel prize for medicine for his work. Seven years later the World Health Organization (WHO) adopted DDT as the main weapon in its campaign to eradicate malaria worldwide. But doubts were

CONTAMINATED CREATURES

DDT can be disseminated over great distances by air and sea currents, while condensation deposits it on glaciers in the polar regions. In 1964 DDT was discovered to be in the fatty tissue of Adelie penguins in the Antarctic and in the livers of polar bears in the Arctic. A study conducted in 2008 showed that penguins living off the krill that thrives in glacial melt water are still contaminated with DDT to this day.

The threat to birds
A normal peregrine falcon's egg (far right) and one affected by DDT (right), which has the effect of thinning the shell.

pyrethroids, carbamates, sulfones and sulfonates. None of these products is entirely anodyne in its effects. Organophosphates are highly toxic, but degrade rapidly. Some organochlorides are suspected of causing hormonal damage, and other insecticides are regularly accused of having similarly noxious side effects. DDT itself has been suspected of causing cancer, although this charge has never been conclusively proven.

already emerging about the effects of the pesticide on the wider environment. DDT kills large numbers of insects, useful as well as harmful, which were then eaten by other creatures. In this way it had entered the food chain and was suspected of being responsible for declining bird populations.

In 1957 an anti-DDT movement sprang up in New York State. Initially the protesters carried little weight, particularly given the economic stakes involved. That situation changed in 1962 with the publication of Rachel Carson's *Silent Spring*, which bitterly attacked the harmful effects of the pesticide. In effect, DDT's chemical stability gave it the attributes of a persistent organic pollutant. It lingered in the environment for decades and had a tendency to build up inside ecosystems. The criticisms were heeded and the WHO abandoned the use of DDT in 1969. Norway and Sweden banned it in 1970. Most other countries followed in the course of the 1970s, although the UK held out until 1984.

Lingering effects

The Stockholm Conference of 2001 called for a total embargo on persistent organic pollutants, but DDT is still used today in many tropical lands. Farmers around the world also rely on the compounds that succeeded it, including synthetic organophosphates and

A NECESSARY EVIL?

In 2006 the World Health Organization decided once more to recommend the use of DDT for house-spraying in malarial regions and also encouraged the distribution of DDT-impregnated mosquito nets. The reason for this reversal of policy was that malaria continues to kill some 3 million people annually around the world, and no other substance is as cheap or effective as DDT at combating the disease.

DDT in Eritrea *Malaria is still a major cause of death in tropical regions, killing a child in Africa every 30 seconds.*

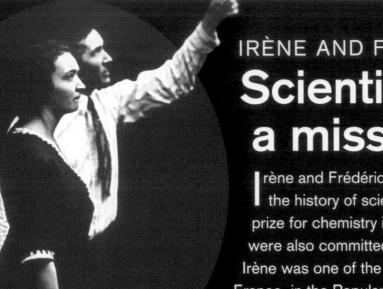

IRÈNE AND FRÉDÉRIC JOLIOT-CURIE
Scientists with a mission

Irène and Frédéric Joliot-Curie had a profound impact on the history of science. They were joint winners of the Nobel prize for chemistry in 1935 for their work on radioactivity. They were also committed humanists who led eventful political lives. Irène was one of the first women to hold a ministerial post in France, in the Popular Front government of 1936. Frédéric was active in the French Resistance and after the war led France's civil nuclear programme, while campaigning against nuclear weapons.

Atomic pioneers
Irène and Frédéric Joliot-Curie belonged to a generation of scientists who focused their research on the structure of the atom. In 1932 their British counterpart James Chadwick, a one-time colleague of Ernest Rutherford, discovered the neutron, one of the constituents of the atomic nucleus (bottom right).

When Frédéric Joliot was appointed to the French Institute of Radium in 1924, at the age of 24, times had changed considerably since the days when Pierre and Marie Curie worked in a makeshift laboratory in a hangar. Marie herself had launched the ultramodern institute, which benefited from generous financial backing when it opened in 1914. It was there that the young man fell in love with Irène, the daughter of the scientific superstar couple. She was three years his senior and had been given the job of training him in radiochemistry. The two married in 1926, united not just by their passion for science, but also by their political beliefs.

A joint Nobel prize
Carrying forward Marie Curie's work, the two were soon making internationally significant contributions to the study of radioactivity. They made advances in research into positive electrons, narrowly missed discovering neutrons (discovered by James Chadwick in 1932) and were responsible in 1934 for identifying artificial radioactivity – the possibility of creating radioactive nuclei that would not naturally exist. This breakthrough, which proved to have many applications and underlies much of nuclear physics, won them the Nobel prize in chemistry in 1935.

Taking a stand
They also found time for active politics. In an era when Frenchwomen still did not have the vote, Irène joined the feminist Committee of

SCHOOL FOR GENIUSES

When Irène was nine years old, her parents Pierre and Marie Curie – who between them won three Nobel prizes for their scientific researches – decided that no available school could provide the intellectual stimulus she needed to nurture her precocious talents. So in 1907 they came to an informal agreement with leading scientist friends, including Paul Langevin and Jean Perrin (himself a Nobel prize winner in 1926), to teach each other's children in their own homes. The 'Co-operative', as the group was known, covered subjects in addition to science, including sculpture and Chinese, and also put great emphasis on play and self-development. One generation on, Irène's own daughter would marry Langevin's grandson.

French Women. Alarmed by the rise of Fascism
in Europe, the couple gave their backing to a
socialist party, the SFIO, and campaigned for
French support for the Republican cause in
the Spanish Civil War (1936-9). When the
Popular Front took power in France in 1936,
Irène was one of three prominent women to
accept posts in the new government. As Under-
Secretary of State for Scientific Research, she
promoted the creation of a state-supported
research organisation, an idea that bore fruit
in 1939 in the form of the National Centre
for Scientific Research (usually known by its
French acronym as the CNRS).

War and resistance

Meanwhile, Frédéric was working on the
concept of uranium-based chain reaction and
the enormous amount of energy it would
release. He also did important work on atomic
reactors, proposing the use of heavy water as
a neutron moderator. Shortly before German
forces reached Paris in 1940, he secretly sent
his stock of heavy water to Britain, along with
all his papers, to prevent them falling into
enemy hands. Irène spent most of the war years
in Switzerland convalescing from tuberculosis,
but Frédéric stayed in Paris, working with the
Resistance. He joined the Communist Party in
1942. He played an active part in the liberation
of the French capital in August 1944.

The post-war years

With the restoration of French national
sovereignty, General de Gaulle charged
Frédéric Joliot-Curie with the task of
reorganising scientific research in the
context of the new republic.
Appointed director of the CNRS,
he helped to transform it into a
major multi-disciplinary body.
In 1945, together with Irène, he
also did much to promote the
creation of France's Atomic Energy
Commission, where Frédéric's
organisational and engineering
talents flowered. The couple opened
France's first nuclear reactor in 1948.

By that time the Cold War was a hard
fact and Frédéric's communist sympathies
began to come under government scrutiny.
A staunch opponent of nuclear weapons, he
was appointed president of the World Peace
Council in 1950, the same year that he was
removed from his post at the Atomic Energy
Commission. He died in 1958, two years after
Irène who, like her mother, had succumbed to
leukaemia contracted as a result of handling
radioactive substances for much of her career.

**Women in
government**
*A magazine cover
shows Irène Joliot-
Curie (right) and
two other women
invited to join the
Popular Front
government in 1936,
despite the fact that
women in France
could not vote.*

Turbojets power a new age of aviation

The Germans won the race to be the first to build one, but Britain took the lead in the development of jet aircraft after the war. Yet it was US interest that assured their long-term future.

German pioneers
Nazi Germany took the lead in the development of jet aircraft, driven by military ambitions. The Heinkel He-178 (above right) served as a model for the He-280 twinjet fighter, but development of this plane was eventually abandoned in favour of the Messerschmitt Me-262 (below). Several versions of the Me-262 were built, including the ultrafast 262 A-2a fighter-bomber, a favourite of Hitler.

It was 27 August, 1939, five days before the German invasion of Poland, and frantic diplomatic efforts were being made to avert war in Europe. On that day a silver silhouette rose above the Marienehe aerodrome in north Germany. What was unusual about it was that there was no propeller at the front of the fuselage or on the wings' leading edge. The plane, a Heinkel He-178, made an unfamiliar whistling sound, which came from its engine, an HeS-3B turbojet. The test pilot, Erich Warsitz, was airborne for just a few minutes, but it was enough to reach a speed of 640km/h (400mph) at a time when the very best fighter planes could manage only 570km/h (350mph). A new era of aviation had dawned.

Race to be first

There were people in Britain who had no intention of letting Germany monopolise jet power. In 1930 an RAF officer named Frank Whittle had taken out a patent on his own turbojet engine. The principle involved went back to Sir Isaac Newton and his Third Law of Motion: there is no action without a concomitant reaction. Air drawn into an engine with sufficient force could produce enough thrust to drive a plane forward. It could do so while employing few moving pieces and at speeds well above the 800km/h (500mph) that represented the theoretical maximum of piston-driven propeller engines. Air Ministry officials were initially sceptical, but eventually gave the green light for the construction of a prototype, the Gloster E28/39, which first flew in May 1941.

Meanwhile, in Germany it took four years from Erich Warsitz's first flight to get jets onto the production line. Heinkel were out of the picture by then, and it was the Junkers company that supplied the Jumo-004B turbojet engine for the Messerschmitt Me-262, the

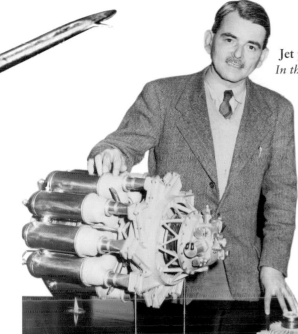

Jet pioneer
In this 1948 photograph, Frank Whittle is seen posing with the WE/700 engine used to power the Gloster E28/39. Whittle was knighted that year by King George VI and given an ex-gratia award of £100,000 for his work on the jet engine.

HOW JET ENGINES WORK

The turbojet that propelled the world's first jet plane was the brainchild of Hans Joachim Pabst von Ohain, a German engineer employed by the Heinkel aircraft company. In the engine, also known as a gas turbine, air entering the nacelle or engine housing was compressed in a compressor before being mixed with fuel and combusted. The violent release of burning fluid through the turbine at the rear drove the aircraft forward. In these original jet engines, all of the air entering the nacelle passed through the motor, but this was not so in turbofan engines introduced in 1960. These not only had fans at their mouths to draw in more air, but also sought to reduce engine noise and fuel consumption by causing part of the airflow to bypass the engine core, being ejected through the exhaust nozzle at the rear, along with the heated gases.

world's first operational jet fighter. When Adolf Galland, head of the Luftwaffe's fighter force, first took the controls of the plane in May 1943, he described the experience as like 'flying on angels' wings'. Another 14 months passed before the twinjet came into service, in relatively small numbers, by which time it was too late to affect the course of the war. Even so, the Allies had reason to fear the newcomer, which could reach 870km/h (540mph) and did much damage to their aircraft.

Post-war lift-off

In 1945 British aircraft engineers took advantage of their own technical advances, as well as Germany's defeat, to take the lead in jet fighter design. Besides the Gloster Meteor, which was used to bring down German V-1 rockets, they could also boast the de Havilland Vampire. British turbojets, based on Whittle's designs but manufactured under licence by General Electric, served to power the US Bell P-59 Airacomet, while the Soviets tried to make up for lost time by buying Rolls-Royce

British jet power
The Gloster E28/39 (above) was the first British jet to fly. It paved the way for the development of the Gloster Meteor fighter (top right), the first 20 of which were delivered to the RAF in June 1944.

BRITISH PIONEER

Born in a terraced house in the Earlsdon district of Coventry, Frank Whittle soon showed himself to be a brilliantly talented youngster with a natural flair for engineering and mathematics. Only 1.52m (5ft) tall, he was initially rejected by the RAF, but eventually passed the physical exam under an assumed name. He patented his pioneering design for a jet engine in January 1930 while working as an Air Force flying instructor. Initially the Air Ministry showed little interest, and it was 1940 before work started on the first British jet prototype, the Gloster E28/39.

JETS IN CONFLICT

The first dogfight between jet fighter planes took place on 8 November, 1950, in the Korean War. The US Shooting Stars had the numerical advantage, but the Russian-made MiG-15s opposing them proved very manoeuvrable and better armed, with one 37mm and two 23mm cannons – weaponry even the F-86 Sabres could not match. On the other hand, the US pilots had the advantage of more accurate gyroscopic sights and above all of greater combat experience than their Chinese counterparts, acquired in aerial combat above Europe and East Asia in the Second World War. By the end of the Korean War in 1953, the Americans had won the air-battle count, with 379 MiGs downed compared to 103 F-86s.

Cold War rivals
Chinese MiG-15s (above) and US F-86 Sabres (below) clashed in the skies above Korea.

Nene engines. By the time that a Meteor set a new world speed record of 975km/h (606mph) in November 1945, it had become obvious that the days of piston-engined fighters were over. The future belonged to jets.

Yet Britain was unable to retain its early lead. France and Sweden both turned out quality competitors, but the main factor was the Cold War, which drove the USA to enter the lists with superior resources. The Bell P-59's successor was the P-80 Shooting Star, the first entirely US-made operational jet fighter, featuring an Allison J33 engine. Making use of German military research, North American Aviation abandoned its plans for a straight-winged fighter in favour of the swept-back design of the F-86 Sabre. The Russians followed suit with the MiG-15,

Turbofan
German scientists worked on turbofan engines in the war years, but the first to go into production was the Rolls-Royce Conway, briefly used in the late 1950s.

which similarly drew on German inspiration. The two rival craft were soon duelling in the skies above Korea.

Before long the performance of turbojets was rivalling that of rocket engines. By the late 1950s the best of them were reaching twice the speed of sound and had combustion systems that allowed thrust to be temporarily increased by injecting fuel via afterburners into the exhaust nozzle. They were designed to climb as high and fast as possible and the prospect of aerial combat was downplayed: guns were largely replaced by air-to-air missiles, and most flights lasted little more than 10 minutes. A decade of research took a high price in human lives: almost one fighter pilot in four died in the cockpit.

Attack and defence

While fighters were increasingly used as interceptors, a very different type of military jet was also making progress: the strategic bomber. In the last months of the war the Luftwaffe had used experimental Arado Ar-234 Blitz twinjets for bombing raids on Britain; their post-war US successors carried first atomic and then hydrogen bombs. In 1946 the US government had set up the Strategic Air Command to do whatever was deemed necessary to ensure national security.

Long range, long run
The B-52 Stratofortress, a long-range strategic bomber, has been in active service with the US Air Force since 1955. The retired B-52s seen here (right) are in storage at the Davis-Monthan Air Force Base in Arizona.

ROCKET ENGINES

Although warplane designers opted for aerobic jet engines that drew in the surrounding air, another technical possibility was open to them: rocket motors, which carried reserves of propellant rather than relying on oxygen from the air to fly. Originally tested by German engineers in 1939 on the Heinkel He-176, rocket engines came back into the picture after the war when the US Air Force became preoccupied with breaking the sound barrier. On 14 October, 1947, above the Mojave Desert in California, test pilot Charles 'Chuck' Yeager became the first man to reach Mach 1 and to unleash the famous supersonic boom, flying a Bell X-1 he had christened *Glamorous Glennis*. The X-1 was powered by a rocket motor fuelled with LOX, a highly explosive mix of alcohol and liquid oxygen that was stored in a tank kept at a temperature of −183°C (−297°F). The rocket plane did not take off on its own, but was carried up to 6,000m (20,000ft) by a Boeing B-29, then released and launched into the air. As it turned out, rocket motors proved to have no great future in aeronautics, although they would play a major part in space flight.

Test drop
A Bell X-1 rocket plane drops away from the bomb bay of the Boeing B-29 launching it on a test flight. The X-1A, with a longer fuselage, reached an altitude of 27,566m (90,440ft) in August 1954.

Fleets of bombers adapted to carry nuclear weapons into enemy territory had to be capable of flying as high and fast as possible to avoid anti-aircraft defences, in particular surface-to-air (SAM) missiles. Jet propulsion met the requirements, but with one major inconvenience: the range of jet planes was insufficient – at least until lowered fuel consumption and the development of in-flight refuelling provided a solution.

The main US warhorse of the skies then became the B-52 Stratofortress, an aircraft of many superlatives. It had a wingspan of 56.4m (185ft) and a gross weight of 221 tonnes, including 22 tonnes of weaponry that could deployed at a range of 16,000km (10,000 miles). Subsequently, every nation with a nuclear deterrent capability felt compelled to equip itself in similar fashion. Over the course of the 1950s and 1960s, though, the improved performance of SAM missiles and of jet interceptors changed the game, forcing bombers to fly below radar cover at very low, rather than high, altitudes. Jet fighters were henceforth able to take on the job, with intercontinental ballistic missiles taking over the main task of deterrence.

Thunderbirds are go!
The US air display
squadron was created in
1953. Today the pilots
fly F-16 Flying Falcons.

VERTICAL TAKE-OFF

For many decades, aeroplanes could only operate from airbases with long runways, which could easily be bombed. In the 1950s aviation designers turned their attention to vertical take-off and landing (VTOL) planes that only needed a space the size of a tennis-court to get airborne. An important breakthrough came with the Rolls-Royce thrust measuring rig, developed by Dr Alan Arnold Griffith, which made use of two turbojet engines mounted back to back horizontally within a steel framework and raised on four legs. With no wings or blades, it was dubbed the 'Flying Bedstead', but it led to the only operational VTOL plane yet manufactured in large numbers: the Hawker Siddeley Harrier and its naval equivalent, the Sea Harrier. A VTOL version of the Lockheed F-35 Lightning stealth fighter is currently in development in the USA.

Riding on a jet plane

The constant improvements made to jet engines had an impact on civil aviation, above all in the USA. Yet even though air transport there had easily outpaced the railways, airlines continued to face economic constraints unknown to the military. The fact was that jets used a lot of fuel and were expensive to maintain. In addition, there was at first a question-mark over their reliability, the more so given that excellent propeller-driven craft like the four-engined Lockheed Constellation and the Douglas DC-6 were already widely available, both of them sound-proofed and provided with pressurised cabins.

EJECTOR SEATS

People began looking for a way of evacuating pilots from aircraft in distress as early as the 1930s. In 1941 German engineers working for the Heinkel company were the first to make a workable ejector seat: the pilot could operate the device in a fraction of a second by pulling a cord between his legs, which caused compressed air to fire the seat out of the stricken craft. In the years after World War II the British Martin-Baker company specialised in producing the seats, but early models sometimes failed to throw the pilots of low-flying planes high enough into the air for the parachute to operate. The firm eventually solved the problem, developing a model with a so-called 'zero-zero' capability that propelled the seat on a sufficiently long upward trajectory, with lowered acceleration to reduce the g-forces operating on the ejectee's spine. Ejector seats subsequently became standard equipment on all combat aircraft.

Against this background, in 1952 the British de Havilland company boldly launched the world's first commercial jet airliner, the Comet, on the London–Johannesburg route. Initially, all went well, but then a series of accidents in 1953 and 1954 forced temporary withdrawal of the planes from service. At the time, the only other jetliner was the Soviet Tupolev Tu-104, but the Comet had shown that wealthy customers liked the speed and comfort of jet travel, and the US Boeing company now entered the frame. Boeing had been forced to lay off many of its workers at the end of the war, but it now expanded its civilian arm. By the time the Comet's problems had been traced to the square shape of its windows, Boeing had developed the 707, which carried twice as many passengers as its British rival. The first Boeing 707 went into service in October 1958 between New York, Paris and London. Passengers were delighted to find journey times cut in half as the planes cruised at 960km/h (600mph). Abruptly, propeller-driven airlines became an anachronism. A transport revolution was under way, and the world had suddenly got much smaller.

Pioneer plane
Passengers board the de Havilland Comet, the world's first commercial jet airliner, for its inaugural flight from London to Johannesburg in 1952. Cruising at 790km/h (490mph), the Comet was half as fast again as its piston-engined competitors.

Workhorse of the skies
A Boeing 707 coming in to land at Heathrow. The plane, which made its maiden flight in 1954, carried passengers around the world for more than five decades – from 1958 to 2010. For two of those decades it was the flagship of the global civil-aviation fleet.

121

Science in the service of war

Science played an important part in the First World War, but in the Second it proved decisive. The image of scientists as people working for the good of humanity would be badly tarnished as their work, if not their intentions, opened the way for a seemingly endless succession of weapons capable of causing mass destruction.

Strike force
British tanks roll into action in the First World War. Built with the encouragement of Winston Churchill, Minister of Munitions from July 1917 on, the new weapons first had a decisive impact at the Battle of Cambrai that November.

The Industrial Revolution transformed the art of war by making the ability to mass-produce arms and supplies a major contributing factor in military success. When the contending parties were of roughly equal economic strength, the capacity for innovation could become crucial: winning became a matter of making faster technological progress.

In this respect, the American Civil War (1861–5) could be said to have been the first modern conflict, witnessing such innovations as machine guns, telegraphs, ironclad battleships, armoured trains and submarines. Some of these also played a part in the Franco-Prussian War of 1870. German industrial might loomed large in this conflict, further augmenting the firepower of German troops with developments, such as breech-loading Krupp cannons and delayed-action shells.

The First World War

The search for a breakthrough in a conflict bogged down in trenches saw chemical laboratories pressed into military service. The Germans producing first tear gas and then, from 1915, poisonous compounds based on chlorine. In response the Allies mobilised their own chemical industries and produced gas masks. In 1917 the Germans introduced mustard gas, against which existing gas masks were ineffective. Poison gases would kill 100,000 people and injure 1.2 million more in the course of the conflict.

Aircraft had hardly been invented before they became weapons of war. The planes available at the time proved relatively ineffective as bombers because of the low payload they could carry, but they were useful for aerial reconnaissance, flying over enemy

lines to photograph troop deployments. As mastery of the skies became a decisive factor in the fighting, anti-aircraft artillery was developed and aerial dogfights became commonplace. Both sides sought to produce faster, more manoeuvrable aircraft. Cockpits were reinforced to protect pilots, while firepower was improved by synchronising machine guns to fire through propellers. In early 1917 the introduction of the Albatros D-III biplane gave the Imperial German Army Air Service a distinct advantage over their Allied opponents, for whom April 1917 would go down in memory as 'Bloody April'. The British-made Sopwith Camel redressed the balance that July; at the end of the war it would be credited with shooting down 1,294 enemy aircraft, more than any other Allied fighter in the conflict.

The machine-gun had forced troops to take cover in trenches. In turn, the Allies sought to end the stalemate of trench warfare by inventing the tank, but the breakthrough was slow in coming. On their first battlefield appearance in 1916 the vehicles performed disastrously, proving slow and difficult to manoeuvre. It was only with the introduction of the British Mark IVs in 1917 and the French FT-17s the following year that tanks began to have a significant impact on the fighting.

In general, the First World War established technological innovation as a significant strategic factor in attaining military success, preparing the way for the decisive role that technology would play in the Second.

Aerial dogfight. *The Fokker D-7 (top right), one of the finest World War I fighter planes, came into service in April 1918. By that time the Germans had already lost the war in the air and the skies were dominated by British Sopwith Camels and SE-5s, seen above bearing RAF roundels.*

War on the airwaves

Control of the airwaves had a significant impact on the Second World War. Radar played a vital role in the Battle of Britain, providing advance warning of enemy raids so that defensive measures could be taken. Similarly, sonar proved crucial in the struggle against U-boats in the Battle of the Atlantic.

Radio too played a central role. Radio telephones were used to coordinate army manoeuvres and special missions over long distances. Besides being indispensable equipment for aircraft and ships at sea, portable radios were used by the army to direct mobile operations that were the more effective because instructions could be modified at any time. The fact that German tanks in Russia had radios, for example did much to make up for their inferiority in most other respects to the Soviet T-34s, which had none. Walkie-talkies (see box, page 124) were

123

Loud and clear
Winston Churchill uses a walkie-talkie in June 1942 (left), while following the progress of a parachute landing at the US Army's Fort Jackson base in South Carolina.

available to the infantry toward the end of the war. Ciphering and decoding messages and instructions became an important part of the ensuing war of the airwaves.

Radio broadcasting was another weapon in the conflict. Hitler and Mussolini used it as an important propaganda tool. The Allies responded with counter-propaganda of their own. The BBC carried Churchill's speeches to all the occupied countries, calling people to arms, while the proclamations of France's General de Gaulle could be heard on the World Service ('Radio London') and Voice of America. Radio London also regularly carried coded messages for the French Resistance.

War in the air

Germany and the Soviet Union took the lead in adapting rockets to carry bombs. German engineers built what was in effect the world's first cruise missile in the unguided V-1, whose straight-line flightpath and relatively slow speed (about 650km/h, or 400 mph) meant that almost half were intercepted. The V-2 was a very different beast: the world's first ballistic missile, it had a range of 300km (185 miles) and struck its target at speeds of 5,600km/h (3,500mph), giving the Allies no chance to counter it. Fortunately, V-2s only became operational in September 1944, and for all the

terrible damage that they did to the civilian populations targeted, they had little effect on the outcome of the war.

At the start of the conflict Germany benefited from a real technological superiority in planes and tanks, which were faster and better armoured than their Allied counterparts, but the Allies soon turned the situation around. For Britain the iconic fighter of the Royal Air Force was the single-seater Spitfire which was key to victory in the Battle of Britain in 1940. First introduced in 1938, its elliptical wing design made it both fast and manoeuvrable. The plane, which was designed by R J Mitchell of Supermarine Aviation (a subsidiary of Vickers-Armstrong) was continually refined and improved throughout the war with over 20,000 being built.

The Germans responded by upping the stakes once more, introducing the Tiger-class tanks, the world's first jet fighter in the shape of the Messerschmitt Me-262 and rocket planes, these last produced too late and in too small numbers to change the course of the fighting. In response, from 1944 on, US B-29 Superfortress bombers with a range of 6,000km (3,700 miles) mercilessly pounded armaments factories and other manufacturing in Germany and Japan. They were backed up by aircraft carriers, first introduced in 1922, which significantly extended the combat range of fighters in the Pacific theatre.

Meanwhile, the invention of nylon in 1938 provided the material for the industrial manufacture of parachutes – some 3.8 million were produced in the course of the war. These

Mobile landing strip *The USS* Yorktown *at anchor in Pearl Harbor. The carrier was sunk in June 1942 at the Battle of Midway.*

ARTIFICIAL INTELLIGENCE AT WAR

Before 1939 engineers performed most of their calculations manually with the aid of slide rules. The war speeded up the development of calculating machines. From 1941 to 1944 the Wehrmacht analysed aircraft performance with the aid of Konrad Zuse's Z3, an advanced programmable calculator that nevertheless remained slow to operate. Meanwhile, in 1942, the US army commissioned the ENIAC computer to work out shell trajectories, but it was 1946 before the machine was ready for use. For Britain the development of calculators, used mainly in the analysis of information, was something of a war aim. Alan Turing developed such machines to decipher first the Enigma and then the Lorenz codes, and in doing to paved the way for the computer age in the post-war years.

Luftwaffe warhorse
The Heinkel He 111 was the principal German bomber at the time of the Battle of Britain (above). The planes were also used to transport supplies and as torpedo bombers in the Battle of the Atlantic.

Computer pioneer
The German engineer Konrad Zuse (1910–95) surveys a replica of his Z1 mechanical calculator, now in Berlin's Museum of Technology. The photograph was taken in 1989.

125

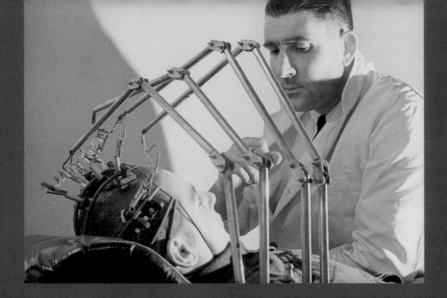

employing the victim's own tissue saved many combatants who had been badly burned. New Zealander Archibald McIndoe, who pioneered the technique while working for the RAF in Britain, was also responsible for advances in plastic surgery, notably in nose reconstruction.

One of the greatest victories of the war was the neutralisation of typhus, which had killed millions of people in the First World War. The vaccine, developed in 1938, only became widely available after 1943. Its efficacy was increased by the use of DDT, which was sprayed on millions of refugees and liberated prisoners to kill the lice responsible for spreading typhus and other disease.

Weapons of mass destruction

The single most significant development of the war was the Manhattan Project, which led to the creation of the atomic bomb; by 1945 some 140,000 people, including thousands of scientists, were involved in the programme. It was the first time in history that a research effort attained such industrial dimensions. Arguably, dropping the bomb saved months, possibly years, of bloody combat on Japanese soil. But even so, the estimated 200,000 people killed in the blasts at Hiroshima and Nagasaki as well as the awful suffering of those struck down with radiation sickness and burns, forced politicians, scientists and people in general to face up to science's dark side.

in turn paved the way for the airlift, which could deliver thousands of infantrymen to a battle zone in record time.

Medical advances

Innovations in medicine were among the positive advances to come out of the war. The most significant was the large-scale production of penicillin; 400,000 doses were produced in the USA in time for the Normandy landings and hundreds of thousands of wounded individuals benefited from them. In other fields, the Dutch physician Willem Kolff made major steps forward in renal dialysis, using cellophane tubes to treat kidney failure that would otherwise have been fatal. Skin grafts

Medicine at war
In December 1940 a doctor in a British military hospital employs an electro-encephalograph, first used in the 1920s, as a diagnostic tool.

The fires of hell
Napalm – shown here (below) in a planned demonstration of its destructive power – was invented in 1942. It is a petroleum-based inflammable liquid designed to stick to objects and to people and to burn at a fixed temperature. In 1980 a UN convention banned its use against civilian populations. The USA never signed the agreement, but claims to have disposed of its stocks in 2001.

BIOLOGICAL AND CHEMICAL WEAPONS

As early as 1933 the Japanese military started testing bacteriological weapons on Chinese prisoners of war. By the time of the Battle of Changde ten years later, they were employing fleas contaminated with bubonic plague to spread infection among the civilian population. Chemistry was also put to use to develop both the cyanide-based Zyklon B gas used in Nazi extermination camps and also nerve gases that were almost impossible to counter because they were absorbed through the air and skin. The terrible trinity of tabun, sarin and soman were produced in large quantities in Germany, even though a dose of just one tenth of a milligramme is enough to kill. Fortunately, they were never used in combat for fear of reprisals.

AN UNINTENDED EFFECT

By persecuting Europe's Jews and denouncing 'Jewish sciences' such as nuclear physics, the Nazi regime inadvertently deprived itself of any chance of developing nuclear arms. Albert Einstein and other scientists, including Edward Teller and Leo Szilard, found refuge in the USA, where they played a leading role in the Manhattan Project.

It was scientists who had provided the means of perpetrating such horrors, and their image suffered as a result. Albert Einstein symbolised the dilemma. In 1939 he had encouraged President Roosevelt to launch research into nuclear arms, but after the war he considered this to have been the greatest mistake of his life. He set up the Emergency Committee of Atomic Scientists, which campaigned to ban the use of nuclear weapons. More realistically, Linus Pauling, winner of the Nobel prize for chemistry in 1954, gathered 11,000 signatures from scientists around the world, helping to limit nuclear tests in 1962 and bringing him a second Nobel, this time the peace prize.

An impossible dividing line

Military advances have often led to benefits in the civilian sphere. Long-range bombers and jet engines, for example, paved the way for

Zemiorka rocket
The Soviet R-7 was the first intercontinental ballistic missile. The early Mars probes shared its bottom-heavy profile, with four liquid rocket boosters attached to the lower part of the fuselage.

long-distance civil aviation, while guided missiles helped to launch the Space Age and with it Earth observation satellites to provide data on the weather. Decoding machines prefigured modern computers. Radar would prove useful to weather forecasters, who employ it to detect bands of precipitation; in time it also triggered the introduction of microwave ovens. Sonar has made maritime navigation safer. Even the technology behind the atomic bomb advanced the development of nuclear reactors to produce energy.

Yet science has also been responsible for the development of thermonuclear bombs and bacteriological weapons, along with napalm and other horrors. Guided missiles and drones, whether used for surveillance or attack, are now promoted as surgical-strike weapons when in fact they continue to cause significant collateral damage – the term the military chooses to use to describe unintended consequences of raids.

Despite the ending of the Cold War, the USA continues to give priority to weapons research, with more than half of its entire scientific research budget going to military uses. The invention of the Global Positioning System (GPS) and the internet owe something to this work, but it remains to be seen whether such benefits will outweigh the role of science in magnifying the destructive force of modern warfare. Since the Second World War, scientific research and innovation have been regarded with a certain ambivalence, and that scepticism looks set to continue.

FROM MISSILES TO SATELLITES

Sergei Korolev, the director of the Soviet programme to build the first intercontinental ballistic missile, was also the man who persuaded the USSR's rulers to put the first satellite into orbit. His engineers went to work with a will, and Sputnik 1 was launched in October 1957. At the time Soviet newspapers reported the story on the inside pages. It was the heavy coverage given to the event in the Western media that convinced the authorities they had achieved a major propaganda coup.

Nuclear fission unchains the mushroom cloud

In 1939 German scientists unleashed a new source of vast energy by splitting uranium atoms. Learning how to control the fission reaction marked a turning point in history. Nuclear power became an inescapable fact, first as a terrifying weapon, then later for peaceful ends.

On 16 January, 1939, two German scientists, Otto Hahn and Fritz Strassman, published the results of a radiochemical experiment. They had established that the heaviest atom, that of uranium, could be split into fragments by the impact of a simple particle, the neutron. Even Hahn and Strassman were surprised by their result – they concluded their paper with the words: 'It is possible that we are mistaken.' A few months later, though, all such doubts were swept away, as it became clear not only that nuclear fission was possible but also that it could release quite phenomenal amounts of energy. So much so, in fact, that Albert Einstein wrote in person to US President Franklin D Roosevelt to warn him of the risks involved if Nazi Germany were to develop an atomic bomb.

Fission made visible

In an ionisation chamber a neutron strikes a uranium nucleus, splitting it into two fragments that will themselves go on to shatter other nuclei, thereby starting a chain reaction.

Protons, electrons and neutrons

Yet how could something as tiny as an atom produce such huge results? The story of splitting the atom got under way in the early 20th century, with the discovery of the atom's internal structure. In 1909 the physicist Ernest Rutherford had described the atom as a planet orbited by satellites. Its nucleus – the planet – consisted of positive electrical charges, the protons, surrounded by negative particles, the electrons. Rutherford's model had limitations that Nils Bohr overcame in 1913 by giving electrons stable, quantised orbits. In 1932 James Chadwick discovered that the nucleus also included a supplementary particle, the neutron. In that same year British physicists John Cockcroft and Ernest Walton actually split an atom – the first instance of the atomic nucleus of one element being changed to that of another by artificial means.

Artificial transmutation

Chadwick's work in showing that the nucleus of an atom was an assemblage of protons and neutrons, concentrated in a very small volume and bound together by gigantic force, was to prove crucial in the history of nuclear fission, because these protons and neutrons could serve as projectiles capable of transforming atoms. In 1933 Frédéric and Irène Joliot-Curie established this fact experimentally: they bombarded aluminium with neutrons and protons and obtained phosphorus and silicon, two different elements. In the course of their transmutation, the aluminium atoms released energy in the form of radioactivity, which previously had been known in its natural state. In effect, the Joliot-Curies had discovered an artificial means of forcing nature to release huge amounts of energy.

What was the mechanism at the heart of the transformations? Italian scientist Enrico Fermi discovered the answer by bombarding uranium with neutrons. The substances obtained disconcerted him – initially he thought he had

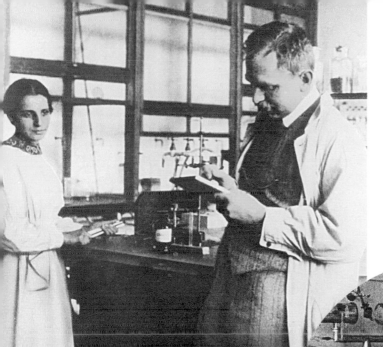

artificially created 'transuranic' elements. But he was mistaken. Without knowing it, he had discovered nuclear fission. The date was 1934.

The theory of fission

When Fermi published his results, Ida Noddack, a German chemist, suggested that the neutrons had shattered the uranium nucleus into fragments, thereby generating lighter elements. The scientific community dismissed the proposal, considering it a rank impossibility that a simple neutron could shatter the heaviest known atomic nucleus. Yet the thought had been sewn. In Berlin Hahn and Strassman decided to repeat Fermi's experiments. They came up with the same result: uranium bombarded by neutrons produced barium, a much lighter element, along with krypton, a rare gas. They published the results early in 1939 and sent the paper to two former colleagues, Lise Meitner and her nephew Otto Frisch, now in exile

Partners in science
Lise Meitner (1878-1968) and Otto Hahn (top left) worked together for almost 30 years and both played a major role in the discovery of nuclear fission, but only Hahn was rewarded with the Nobel prize in chemistry in 1944. In contrast, Frédéric and Irène Joliot-Curie – pictured here (left) soon after their discovery of artificial radioactivity in 1934 – were jointly awarded the highest scientific honour.

due to the Third Reich's anti-Jewish policies. Seeing the significance of the experiment, Meitner and Frisch came up with the idea of fission that Noddack had presaged. When struck by a neutron, they claimed, the uranium nucleus becomes unstable, not being able to accept a supplementary particle. It loses its shape and splits into two fragments of roughly equal mass. When it shatters, some of the energy binding the protons and neutrons within the nucleus is released. Meitner worked out the amount of energy using Einstein's $E = mc^2$ equation: 200 billion electron volts per atom.

Frisch communicated the results to Nils Bohr, who was about to embark for the USA. En route, Bohr stopped off in London, where he received a welcome that was no more than polite; war with Germany was looming and no-one, it seemed, had the time to concern themselves with the chimera of atomic energy. So he duly passed on to America.

Chadwick's detector
James Chadwick built this detector apparatus, equipped with an electronic amplifier, to establish the existence of neutrons.

A NEAR MISS

If Hahn and Strassmann had used pure uranium 235 in their experiment, they would have set off a chain reaction that would have caused an explosion big enough to wipe out not just themselves but the entire city of Berlin.

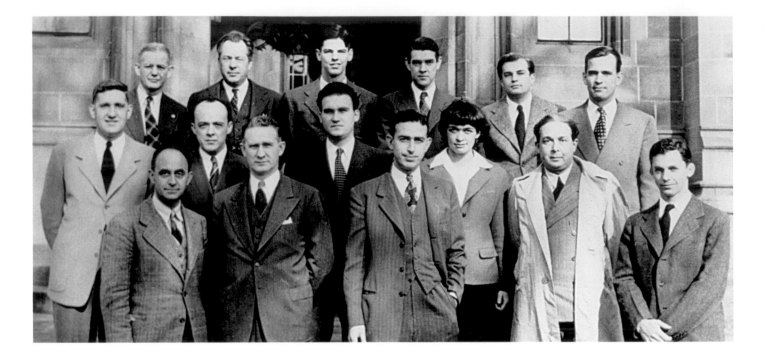

The team responsible for creating the world's first nuclear reactor gather in December 1946 for a reunion at the University of Chicago. Enrico Fermi is on the far left of the front row; Leo Szilard, wearing a coat, is second from the right.

Controlling chain reactions

A new race now got under way. The challenge was to prove that fission could take place in a solid mass of uranium by a process of chain reaction – splitting a nucleus would release at least one neutron that would then set another fission in motion, and so on. In this way a single gramme of uranium could unleash as much energy as 3 tonnes of coal. Two different scenarios presented themselves: that of a controlled reaction serving to produce energy that could be put to use, or alternatively an uncontrolled reaction that could generate a weapon of mass destruction. In May 1939 Joliot and his colleagues took out patents on the production of energy from uranium.

In the USA, meanwhile, a physicist named Leo Szilard, who had worked with Fermi on chain reactions, was growing concerned that Nazi Germany might be the first country to develop an atomic bomb. He communicated his fears to President Roosevelt, using Albert Einstein as an intermediary. On 2 August, 1939, Einstein wrote a historic letter to the

US President, claiming that: 'A single bomb of this type, carried by boat and exploded in a port, might very well destroy the whole port together with some of the surrounding territory.' Roosevelt was initially sceptical, but largely as a result of intense urging by British scientists, who calculated that the Germans were capable of producing such a weapon, he authorised the establishment of an Advisory Committee on Uranium. Fermi and Szilard were able to work on the development of a first nuclear reactor, although with limited funds: their $6,000 budget largely went on buying 50 tonnes of uranium oxide along with 4 tonnes of graphite.

The battle for heavy water

The graphite was essential if the chain reaction was to unfurl smoothly, for it served as a moderator slowing the speed of the neutrons released by the multiple fissions. When a uranium nucleus splits, the neutrons that are released travel at roughly 20,000km/s (12,500 miles per second). They need to reach the other uranium nuclei at a much reduced pace if they are to be absorbed and provoke fresh fissions.

There was another moderator that was even more efficient than graphite, but it was also more difficult to obtain. This was heavy water, which became the subject of open conflict in 1939. The only manufacturer of the substance was the Norwegian company Norsk Hydro. German agents tried to buy up the company's stock, but a French delegation got there first – just before Nazi forces invaded and occupied Norway in April 1940. The Allies then staged bombing and commando

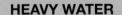

HEAVY WATER

Heavy water is chemically close to plain water (H_2O), except that it is enriched with deuterium, an element whose nuclei contain neutrons whereas hydrogen has none. Heavy water effectively slows the neutrons responsible for fission, but is difficult to produce, requiring a sequence of different chemical procedures. The hermetically sealed bottle (left) contains 25g (just under an ounce).

raids to prevent the occupation forces from restoring production of the precious fluid.

Atomic research hots up

Meanwhile, research was continuing in the USA. In 1939 Nils Bohr discovered that, of the three forms of uranium found in the natural state, only one was suitable for fission: uranium 235, with a nucleus composed of 235 protons and neutrons. This form was also rare, so the Advisory Committee on Uranium set out to find a way of obtaining it from the much more abundant uranium 238.

By this time, Fermi had managed to carry out a controlled chain reaction. In December 1942, his atomic pile – in effect, the world's first nuclear reactor, installed in a racquets court under the stand of an abandoned sports stadium at the University of Chicago –

produced some hundreds of watts of energy for a period of 28 minutes. The success of the experiment confirmed the strategy adopted by the US authorities, who were completely reorganising atomic research. The outcome was the Manhattan Project, an extraordinary research programme that managed to remain highly secret while mobilising more than 140,000 people around the nation.

Three years, two bombs

Given a vast budget and placed under military control, the Manhattan Project provided the USA with atomic weapons in three years. It functioned from three main bases: Oak Ridge in Tennessee, where the first factory for the production of uranium 235 was installed; Hanford in Washington State, which produced plutonium, an artificial element thought to be

A CRISIS OF CONSCIENCE FOR SCIENCE

When Albert Einstein wrote to President Franklin D Roosevelt in 1939, seeking to persuade him to launch a nuclear research programme, he only envisaged atomic power being used as a deterrent to Nazi Germany. He would later claim: 'If I had known that the Germans would not succeed in constructing the bomb, I would never have lifted a finger.' In fact the Nazis never really came close to creating the bomb. Claims have been made that Werner Heisenberg, who directed the German nuclear weapons programme, deliberately sought to sabotage, or at least slow down, its progress, but the point remains controversial.

A last look *Einstein and Leo Szilard (above) review the letter to President Roosevelt that helped to inspire the Manhattan Project, which gave the USA atomic weapons.*

Post-war disarmament *In May 1945 Allied forces dismantle an experimental nuclear reactor installed in a secret 'atomic cellar' in the German town of Haigerloch (left).*

Desert test site
The first US nuclear explosion took place at Alamogordo in the New Mexico desert on 16 July, 1945 (below). Dropped from a 30m (100ft) tower, the bomb left a crater half a mile wide and vitrified the sand over a range of a mile.

even more fissile that eventually found its way into one of the first two atom bombs; and Los Alamos in New Mexico, where plans for the making of the bombs were drawn up. Hundreds of physicists worked there, including more than 20 past or future Nobel prize-winners. In charge was Robert Oppenheimer, often considered the father of the atom bomb.

The horror of Hiroshima

By the summer of 1945 two bombs were ready: one, employing uranium 235, was fully operational; the second, plutonium-based, had been tested. On 16 July, 1945, the world's first

NUCLEAR POWERS

The Soviet Union was able to use information obtained from spies who infiltrated the Manhattan Project to speed up the production of its own atomic bomb. A first weapon was tested in Kazakhstan in August 1949. Today five nations are classed as officially recognised nuclear powers, covered by the Nuclear Non-Proliferation Treaty: the USA, the UK, France, Russia and China. India, Pakistan and North Korea have also declared themselves to have nuclear weapons, while Israel refuses to confirm or deny reports that it has developed a nuclear capability.

atomic explosion took place in the desert of New Mexico, releasing a force equivalent to 20,000 tonnes of TNT. Following the capitulation of Germany, some of the scientists involved expressed their opposition to the use of the bombs on Japan, but their voices went unheeded. President Truman authorised the dropping of the 'Little Boy' uranium bomb on Hiroshima on 6 August, 1945, and of 'Fat Man' on Nagasaki three days later. The two towns were wiped out. In Hiroshima more than 78,000 people were killed and 9,500 wounded; Nagasaki saw 36,000 dead and 40,000 wounded. Radiation sickness subsequently affected large numbers of survivors, causing cancers and leukemia.

Following these events, many physicists lent their voices to the anti-nuclear movement; the Campaign for Nuclear Disarmament (CND), founded in Britain, held its first protest march from London to Aldermaston in 1958. Others did their best to reconcile a shaken world to nuclear power by promoting its peaceful use as a source of low-cost energy. And some went on to develop the even more powerful hydrogen bomb, launched in 1952.

The destruction of Hiroshima
Dropped at 8.13am on 6 August, 1945, the 'Little Boy' bomb exploded 530m (1,740ft) above ground (left), raising the temperature almost instantly to 6,000°C (10,800°F). Of the town's 90,000 buildings, 62,000 were totally destroyed (below).

The Lascaux caves 1940

Artistic wonder
Experts examine the Lascaux paintings sometime in the early 1950s (above). Top: The Hall of the Bulls.

On Thursday, 12 September, 1940, 17-year-old Marcel Ravidat set out to explore a hole caused by the recent fall of a pine tree on a hill outside Lascaux in the Dordogne region of France. En route he happened to meet up with three of his friends who decided to go along with him. The four teenagers slipped gingerly through the narrow opening that led underground and found themselves in a large cave. By the light of their torches they gazed in astonishment at the cave walls which were painted with multi-coloured images of animals – bulls, horses, deer.

A REPLICA LASCAUX

In 1948 the entrance to the caves was modified to allow entry to the public. The initiative proved so popular, attracting up to 1,200 visitors a day, it ended up endangering the paintings as they were affected by the higher levels of carbon dioxide that accumulated underground. Another effect of exposure was to encourage the growth of fungi and moulds, causing green patches to appear on the walls. As a result, Lascaux was closed to the public in 1963. Twenty years later, a replica of the two best-known halls was created nearby, giving visitors a fresh chance to admire the beauty of these extraordinary paintings.

Unexpected expertise

A few days later the Abbé Breuil, a specialist in prehistoric cave art, came to investigate. He not only confirmed the authenticity of the paintings but also made the first inventory of the images, which revealed previously unexpected pictorial skills among Stone Age artists. Subsequent research would also suggest a surprising level of technical expertise: the artists heated mineral pigments to obtain the desired colours and used scaffolding to reach the caves' upper levels. Animal-fat lamps, spear points and pieces of cord were also found on the site. Carbon-14 dating carried out in the late 1950s showed that the paintings dated from 17,000 to 20,600 years ago.

With 1,900 separate paintings and engravings along 250m of galleries, Lascaux thoroughly deserves its reputation as the Sistine Chapel of prehistoric cave art. Experts still argue over the exact significance of the images: were they inspired by shamanism or totemism, or were they perhaps some form of hunting magic? What is certain is that in the subterranean gloom artists of the earliest times recorded their myths and beliefs with style and brio. The caves were declared a UNESCO World Heritage Site in 1979.

Scuba diving gear 1943

The term SCUBA – an acronym for Self-Contained Underwater Breathing Apparatus – was coined in the Second World War for the oxygen 'rebreathers' developed by Dr Christian Lambertsen, an army specialist in diving physiology, and used by US frogmen in underwater warfare. Before 1943 divers had to rely on a tube linked to the surface to breathe. The Rouquayrol-Danayrouze apparatus, invented in 1865, included a small air tank, but had only enough air for 30 minutes underwater. In 1933 Yves le Prieur designed equipment that offered twice that, but by providing a continuous flow of air it used up the available supply more quickly than was needed.

A controlled supply

In 1942 Jacques-Yves Cousteau shot the first French underwater film, *18 Metres Down*, without breathing apparatus. The film was a hit, encouraging Cousteau to look for an easier way of filming underwater. He teamed up with an engineer named Émile Gagnan and in 1943 they developed a self-contained breathing kit they called the 'aqua-lung'. The device featured cylinders of compressed air along with an automatic regulator that supplied air on demand. Cylinders carried on the diver's back fed a valve that controlled the airflow, delivering it only when the diver breathed in. Two rubber tubes carried the air from the valve to mouthpieces held between the teeth.

Like a fish in water

'I swim almost effortlessly ... like the fish I meet', Cousteau claimed. Industrial production of the aqualung began in 1946, and proved sufficiently successful to provide for all the inventors' needs thereafter. The demand valve, which is still in use to this day, made scuba diving a popular pastime around the world. And besides its leisure applications, scuba equipment is also used for sea rescues and underwater archaeology.

Free diving

The joint inventor of the scuba set, Commander Jacques-Yves Cousteau (top right) poses with the American diver Terry Young at San Pedro in California in 1950. Both are equipped with aqualungs. The result of a lengthy process of development, the kits made it possible to explore the underwater world with a minimum of inconvenience.

THE FIRST FLIPPERS

The inventor Benjamin Franklin made the first flippers, out of wood, in 1718 at the age of 12, but the modern versions were the brainchild of Louis de Corlieu, a French naval officer who in 1933 made them from rubber strengthened by steel blades. Successfully marketed in the USA, flippers played a part in the Normandy landings on the feet of British commandos and American marines.

temperature separating the cold air in high latitudes from the warmer air at lower latitudes; the other is the subtropical jet, situated at roughly 30° latitude. They occur where air masses of different temperatures meet, flowing not directly from the hot to the cold spots but rather around the boundary of the two masses, deflected by the Coriolis effect.

The stronger the jet streams, the more they risk becoming unstable, swirling and eddying to cause depressions at lower altitudes and also triggering local developments that can include high winds and storms. Jet stream movements can also have long-lasting meteorological effects: Britain's Met Office blamed the unusually wet summers of 2007, 2008 and 2009 on a southward shift of the polar jet stream, blocking off high-pressure weather systems arriving from southern Europe.

Jet streams 1944

Tracks in the sky
A photograph taken from the Gemini 12 space capsule shows a band of cloud carried along by a jet stream above Egypt and the Red Sea (top).

At the end of the Second World War, pilots of B-29 bombers flying over Japan at high altitudes found that they could not release their bombs accurately because the aircrafts' speed was greater than expected. They claimed it was as though they were being carried along on some unknown 'river of air'.

The phenomena responsible were soon identified as jet streams, powerful aerial currents that form at altitudes between 8,000 and 12,000m (26,000–40,000ft) in the tropopause, the uppermost layer of the troposphere. The streams move at an average speed of 200km/h (125mph), with occasional spikes reaching up to 450km/h (280mph). They move around the globe from west to east in the form of long, flattened tunnels that are several thousand miles long and can be hundreds of miles wide, but are typically just 2 to 3 miles deep.

Currents in the sky
Whether in the northern or southern hemisphere, there are two principal, quite separate, jet streams: one is the polar jet, resulting from the marked contrast in

A helping hand for aircraft
Long-distance balloonists seeking to circumnavigate the globe generally aim to catch a ride within jet streams to achieve their goal. Air routes avoid them when flying westward, against the flow, but on eastbound flights take advantage of them to shorten flight times and reduce fuel consumption, despite the violent turbulence sometimes found around their edges. The prevalent direction of jet streams does much to explain why westward flights from London to New York generally take almost an hour longer than east-bound flights from New York to London.

AN UNNOTICED DISCOVERY

In 1926 a Japanese meteorologist named Oishi Wasaburo demonstrated the existence of jet streams using data from weather balloons. Hoping to reach a global audience, he published his findings in Esperanto, but for that reason they remained largely ignored by the scientific community.

Ectothermy 1944

Some desert lizards lie in the sun to warm up and take to the shade to cool down. They do so to keep their body temperature as even as possible, despite external variation in temperature over the course of the day. This fact was established in 1944 by Raymond Bridgman Cowles and his pupil Charles Mitchill Bogert, two American zoologists specialising in reptiles and amphibians.

The discovery might seem obvious today, but at the time scientists still divided the animal kingdom into warm-blooded creatures (homeotherms) maintaining a constant average body temperature, and cold-blooded animals (poikilotherms), whose temperatures varied. The lizard Cowles and Bogert had studied disrupted this neat schema, as it was at one and the same time a poikilotherm, like all other reptiles, and a homeotherm, capable of maintaining a stable temperature.

Humans are endotherms whose temperature rarely strays far from 37°C (98.5°F); breathing and feeding both play a part in regulating this temperature. In contrast, ectotherms may alter their temperature by as much as 20°C (36°F) without putting themselves at risk, with the paradoxical result that so-called 'cold-blooded' animals sometimes have higher temperatures than warm-blooded ones. Like the desert lizards, they are dependent on their environment but have no need to expend energy to adapt to it.

Ectotherm and endotherm
An infrared photograph of a mouse and a snake reveals the difference in temperature between the two creatures, with dark colours representing low temperatures.

A biological thermometer

Emboldened by this discovery, the herpetologists put forward two new concepts. They used the word 'endotherm' to describe species that regulated their body temperature through internal processes and 'ectotherm' for those that relied on the external environment.

Desert dweller
The zebra-tailed lizard (right), taxonomically listed as Callisaurus draconoides, *lives in arid desert regions of the southwestern USA. Capable of enduring great heat, the lizards burrow into the sand at night to keep warm.*

THE REPTILES

Tortoises, lizards, snakes, crocodiles: the reptiles, long referred to misleadingly as 'cold-blooded' creatures, are most easily defined by their differences from other classes of animals. They are not mammals, being without mammary glands with which to feed their young, and they have no feathers, as do birds. There are about 7,000 species in all, most of them covered in scales.

Using vibrations to heat up food

As a result of a chance observation by a military engineer, a microwave radar device developed for anti-aircraft defence led to the birth of a much more peaceful invention: an oven that, after many improvements, found its way into most kitchens.

The Second World War demonstrated the vital strategic importance of anti-aircraft defence. In its wake, Percy Spencer, an engineer with the US Raytheon arms company, was assigned to work on the magnetron, a radar device operating on very short wavelengths designed to detect aerial attack. One day he noticed a pleasant smell that he soon traced to a bar of chocolate in his pocket that was melting.

Some of his fellow laboratory technicians had in fact already found they could use the heat the device gave off to warm their hands or even their meals. Spencer, a born experimenter, put kernels of maize near the magnetron and ended up with popcorn. An egg placed in front of the machine emerged hard-boiled.

A new kind of oven

Spencer quickly grasped the implications of the discovery. To get the most out of the energy the microwaves released, he constructed a metal box that served to contain them within reflecting walls – in effect, the first microwave oven. The temperature of foodstuffs placed inside rose quickly as the waves got to work, first and foremost on the molecules of water that they contained and then to a lesser extent on the sugars and fats. Pulled by the oscillations of the electromagnetic field, the water molecules behaved like tiny magnets, orienting themselves first one way and then the other, and the very rapid vibrations involved generated heat that was spread by conduction. Yet the waves had little or no effect on glass or porcelain containers, which were heated only by the food they contained.

Different models for different needs

An experimental microwave oven was installed soon after in a Boston restaurant. A first commercial model went on sale in 1947, but it took up a lot of space and required special plumbing, as it was water-cooled. It also cost as much as a new car. Unsurprisingly, it was not an overnight success.

Over the years the ovens became smaller and lighter. Now air-cooled, they were provided with a waveguide – a metal propeller that spread the waves, encouraging more even heating. Gradually the ovens were taken up by the food industry and a few big restaurants. During the 1960s their price

DANGEROUS TO USE?

The first microwaves were suspected of causing cancers. Research carried out since has shown that accidental exposure to very high levels of microwaves in industrial settings can cause burns or eye problems, primarily through the heat generated. Microwaves have not been proven to play a part in neurological, hormonal or immunological conditions. Modern domestic appliances are hermetically sealed, so as long as they are properly maintained, they expose their users only to very weak levels of radiation, comparable to those given off by magnetic fields – which are themselves still under investigation.

Getting smaller
The size of microwave ovens has steadily decreased over the years, making them more suitable to a domestic setting. The Raytheon Radarange, the first microwave oven on the market, stood 1.8m (6ft) high. This model (right), from the Tappan Stove Company, dates from 1955.

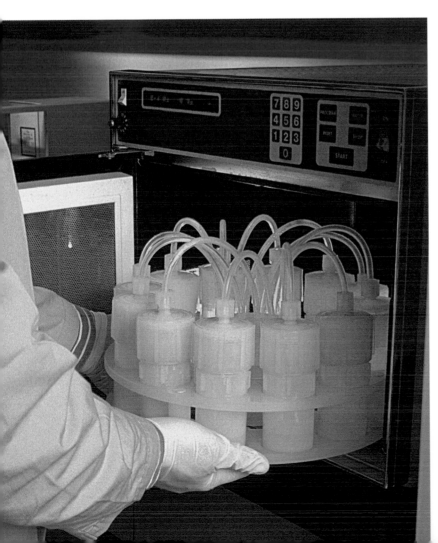

dropped by 80 per cent, and by the 1970s they were a feature of many households; 85 per cent of British households would have one by 2001. The ovens proved a versatile tool, well suited to modern lifestyles; in particular, they could be used to rapidly defrost and reheat frozen foods. Industry also found uses for microwave installations, notably for drying cork, paper, leather and tobacco. In 2007 the Karlsruhe Institute of Technology in Germany equipped itself with the largest in the world, a gigantic oven used in the manufacture of composite materials reinforced with carbon fibre for use in the motor and aviation industries.

Research tool
Microwave ovens are not just domestic appliances. In laboratories they are used to mineralise, dry or evaporate small to medium-sized samples (left).

A modern metropolis

In 1925 New York was the most magnetic and surprising city in America, a centre of the arts, intellectual life and high finance. It was badly hit by the Wall Street Crash of 1929, but soon found its feet again, bouncing back to dazzle the world.

City of dreams
Travellers get a first view of the towers of Manhattan as they arrive in New York by ship in the 1930s (above). The main image of the city skyline (top right) was prepared for the Edison Pavilion at the New York World's Fair of 1939.

New York's harbour had been the place of entry into the USA for generations of immigrants since the 1820s, and its klaxons continued to welcome boats arriving from the Old World well into the 20th century. The city's great illuminated avenues, crisscrossed by limousines bearing the legendary logos of Cadillac and Chevrolet, echoed to the sound of police and fire service sirens. An air link was established between New York and San Francisco in 1920, and Newark Liberty International Airport opened in 1928. The year before, Charles Lindbergh had received a hero's welcome on Broadway when he returned from his trans-Atlantic flight.

Millionaires vied with one another to build the tallest skyscrapers. The Flat Iron Building was the first, on the corner of Broadway and 5th Avenue, at 87m (285ft), rapidly overtaken by the flamboyantly Gothic Woolworth Building reaching 241m (790ft). In 1930 the Chrysler Building briefly held the crown of world's tallest building at 319m (1,047ft) including its spire. But John J Raskob, a former director of General Motors, soared past it. The Empire State Building, which he

commissioned, opened its doors on 1 May, 1931. Measuring all of 443m (1,454ft) to its topmost pinnacle, it remained the world's tallest building until the early 1970s.

The capital of capital

The 'vertical city', as the architect Le Corbusier called New York, was the concrete symbol of the financial success of a nation that daily extended its leadership status in the world economy. The great metropolis owed its extraordinary prosperity first and foremost to its geographic location on the east coast. Built on a natural harbour at the mouth of the Hudson River, its port facilities enabled New York to replace Chicago from the start of the 20th century as the commercial hub of the nation. Two-thirds of all US imports came in through its docks and the city was surrounded by an industrial belt where many leading companies had their headquarters.

With its flourishing banks, New York usurped London's 19th-century position as the capital of capitalism and became the world's showcase of wealth and commerce. 'Everyone ought to be rich!' proclaimed John Raskob, conveniently forgetting the harshness of life in the city's factories and slum districts such as Harlem. Once one of Manhattan's most sought-after residential areas, Harlem was transformed into a black ghetto following the arrival of large numbers of Afro-American migrants seeking work, mostly from the southern states.

Gateway city

More than any other American city, New York came to symbolise the USA to the rest of the world. Individuals wanting to make a success of their lives, or to start again with a clean slate, flocked there from all parts of the globe. The metropolis absorbed waves of emigrants from Europe, more than doubling the population in 30 years, up from 3 million in 1900 to 7 million by 1930. In the process New York became the world's most cosmopolitan city. In 1925 it was reckoned to have the second-largest number of German speakers of

THE DREAM TRAIN

One of the best ways of arriving in New York was on the *20th Century Limited*, a luxury express train covering the journey from Chicago to New York in 16 hours. The washrooms in the sleeping compartments offered a choice of fresh or salt water and manicure facilities for ladies. There were secretarial services for businessmen, delivered by PAs with Hollywood smiles. The train made its way right to the heart of New York through a tunnel that opened onto the world's biggest station, Grand Central, with 68 platforms and a décor rich in marble, copper and gold leaf.

Art Deco town
Veneered in rare woods, the lift door of the Chrysler building (left) leaves no doubt of the loving care that went into the interior decoration of the 77-storey skyscraper, completed in 1930.

In the decades that followed, specifically American styles would blossom, often inspired by artists with European roots. Pop Art flourished under the influence of Andy Warhol, the son of Slovak immigrants, and of Stockholm-born Claes Oldenburg. Abstract Expressionism was shaped by Mark Rothko, originally from Russia, and the Armenian Arshile Gorky. And not just painters but also musicians and film-makers flocked to the USA and its greatest attraction: New York.

Black Thursday

The heady 1920s came to an abrupt end on Thursday, 24 October, 1929. That day the Wall Street stockmarket, which had soared to dizzy speculative heights not backed up by similar progress in the real economy, first wavered and then plummeted. Small shareholders panicked, selling at any price: in all, 13 million transactions took place on that one day. The crisis forced banks, businesses and overstretched individuals into bankruptcy. New York was struck to the heart, its poor facing untold misery and the middle classes also being badly affected. In March 1930, some 35,000 people took to the city's streets to demonstrate and express their anger. Unemployment rose to alarming proportions: by 1933

Stockmarket panic
Crowds gather in Wall Street outside the New York Stock Exchange on 'Black Thursday', 1929.

Abstract pattern
A New York apartment building featuring a typical zig-zag fire escape.

any city outside Germany, and trade union proclamations were printed in both English and German. The Italian community found refuge in a neighbourhood that came to be known as Little Italy. The Irish contributed a disproportionate percentage of the city's police force. In addition, New York had the biggest Jewish population of any city in the world.

Cultural ferment

Among the great mass of immigrants were a number of individuals of outstanding creative talent. The city had discovered European avant-garde art with the Armory Show, a celebrated exhibition of contemporary painting organised by the photographer Alfred Stieglitz in 1913. Collectors vied to buy up modernist treasures ranging from the works of the Impressionists through Art Nouveau and Jugendstil to Cubism and the Italian and Russian Futurists. In 1929 MOMA – the Museum of Modern Art – opened its doors in the heart of Manhattan and rapidly became a world showcase of artistic creativity.

Yet New York was to become much more than just a receptacle for art created elsewhere.

THE BIG APPLE

New York's best-known nickname had been in use for decades before it became known around the world following a publicity campaign in the 1970s. There is some controversy about its origins, but it seems to have emerged in the 1920s in horse-racing circles. One story has it that New Orleans stable-hands used it when dispatching thoroughbreds to New York racetracks. An alternative story traces the name back to an expression popular with jazz musicians and other entertainers: 'There are many apples on the tree but only one Big Apple.' In the 1930s, at a time when people without jobs were reduced to selling apples in the street to earn a little money (right), *The Big Apple* became a popular dance hit for the Tommy Dorsey Orchestra. Wherever it first came from, the expression is now used the world over to describe the city's attractions – but New Yorkers themselves never use it.

Swing time
Glenn Miller and his orchestra performing in a New York nightclub in about 1941.

more than 12 million people were out of work across the USA, representing a quarter of the labour force, while 2 million were homeless.

Recovery through war

Franklin D Roosevelt's New Deal, introduced in 1933, went some way to restoring confidence, but it was the Second World War that finally put the US economy back on track and New York gradually recovered its old prosperity. The Victory Programme, launched in 1942, restored full employment. A massive rise in the number of women workers helped to compensate for the mobilisation of young men to fight abroad; the new recruits found jobs in armament factories making Liberty ships and warplanes. The port of New York was requisitioned for troop transports. New inventions found industrial uses: nylon provided the material for vast numbers of parachute canopies, and plastics, polystyrene and silicon were all exploited to meet military needs. Calculators put in a first appearance.

In 1942 the Manhattan Project – the very name was symbolic – called on the services of the finest physicists of the day. If Albert Einstein, Enrico Fermi, Leo Szilard, Lise Meitner and others had not emigrated to New York in the 1930s, driven there to escape Nazi persecution, the atomic bomb might never have been built. In 1945 New York could boast of being one of the few metropolises around the world not to emerge ruined from the war.

THE GREAT WHITE WAY

New York's entertainment hub centred on the theatres of Broadway and Times Square in the heart of Manhattan. Cinema entered a golden age with the arrival of the talkies in 1927; vast audiences thrilled to the musicals of Busby Berkeley, including *42nd Street* and *Gold-Diggers of 1933*. Blockbusters like *The Wizard of Oz* and *Gone with the Wind* were projected on increasingly large screens. Mickey Mouse himself first spoke at the Colony Theatre in November 1928, which premiered the Walt Disney cartoon *Steamboat Willie*.

GI exodus
Many thousands of soldiers passed through the port of New York after the USA entered World War II in December 1941.

BRIG. GEN. Wᵐ E.H

CHRONOLOGY

The timeline on the following pages outlines
key discoveries and inventions of the hugely
formative years up to the end of the Second
World War. Selected historical landmarks
are included to provide chronological context
for the scientific, technological and other
innovations listed below them.

1914

EVENTS

- The assassination of Archduke Franz Ferdinand, heir to the Austro-Hungarian throne, at Sarajevo triggers the outbreak of the First World War (1914)
- Battle of the Somme (1916)
- Russian Revolution (1917)
- Armistice between the Allies and Germany brings the fighting to a close (1918)

INVENTIONS

- Alfred Wegener publishes *The Origin of Continents and Oceans*, outlining the theory of continental drift that underlies current plate tectonics

- Germany's Hugo Junkers constructs the J-1, the first aeroplane to be made entirely out of metal

- Jay MacLean, a medical student at Johns Hopkins University, discovers heparin, a molecule that will be used as an anticoagulant

- British forces introduce tanks at the Battle of the Somme

- German forces in Flanders deploy mustard gas, fired in shells against the Allied lines

- The Royal Engineers and the RAF pioneer Britain's first airmail service, delivering post to British troops stationed in Germany

- Piggly Wiggly, the world's first self-service grocery store, opens in Memphis, Tennessee, prefiguring modern supermarkets

▲ British SE-5 fighter aircraft of the First World War

1919

EVENTS

- The Treaty of Versailles imposes disarmament on Germany and establishes the League of Nations (1919)
- By the terms of the Treaty of St-Germain-en-Laye, the Austro-Hungarian Empire is dismantled (1919)
- The Treaty of Sèvres puts an end to the Ottoman Empire and proposes Kurdish and Armenian states (1919)

INVENTIONS

- Hugo Junkers introduces the F-13, the first all-metal transport aircraft

- Deutsche Aero Lloyd, a precursor of the German airline Lufthansa, offers the first commercial flight to paying passengers, between Berlin and Weimar

- British pilots John Alcock and Arthur Brown make the first non-stop flight across the Atlantic Ocean in a converted Vickers-Vimy bomber

- Sigmund Freud lays the foundations of psychoanalysis

- The world's first motorway is built in south-west Berlin

- Canadian John Larson devises the first effective lie detector

- Frenchman Louis Rustin invents the bicycle tyre repair patch

- The first gliding competitions are held in Europe, setting new records for unpowered flight

▲ German Fokker D-VII

► A scene from Fritz Lang's film *The Woman in the Moon*

▼ British tanks in the First World War

1923

- Mustafa Kemal Atatürk becomes first president of a newly created Turkish Republic (1923)
- Tangier in Morocco is declared an international zone (1923)
- The Indian Citizenship Act confers full rights of citizenship on native North Americans (1924)
- Death of Lenin (1924)

- In the USA James Cummings and Earl McLeod construct the first bulldozer

- The Kimberly-Clark company in the USA launches Kleenex, the first paper handkerchief

- Hans Berger, a German psychiatrist, uses electro-encephalography (EEG) techniques to study the human brain

- In France the physicist Louis de Broglie proposes the concept of particle-wave duality

1925

- Benito Mussolini establishes a fascist dictatorship in Italy (1925)
- Chiang Kai-shek becomes head of the Kuomintang, China's nationalist party (1925)
- Germany joins the League of Nations (1926)
- Hirohito becomes the 124th Emperor of Japan (1926)

- US physicist Robert Goddard successfully tests the world's first liquid-propellant rocket

- Erik Rotheim of Norway takes out a patent on an aerosol spray

- In the USA Eldec, an industrial cleaning company, produces the first electric steam iron

- French engineers Jean-Albert Grégoire and Pierre Fenaille design the first front-wheel drive car

- John Logie Baird transmits the first television images

- Erwin Schrödinger publishes a series of ground-breaking articles on quantum theory, a revolutionary development in science

▶ An early steam iron in use

▼ One of the first televised images transmitted by Baird in 1926

▼ John Logie Baird's televisor

1927

EVENTS

- The Cristero War breaks out in Mexico, pitting Catholic partisans against an anticlerical government (1927)
- The League Against Imperialism is founded in Brussels (1927)
- Chiang Kai-shek purges Communists from the Kuomintang and defeats northern warlords to reunify China (1928)
- First Soviet Five Year Plan (1928)

INVENTIONS

- The first videophone call is made from New York to Washington DC

- With the help of his brother Cecil and another associate, Louis Agassiz Shaw, the American industrial hygienist Philip Drinker invents the iron lung

- Hans Wilsdorf, founder of the Rolex company, makes the Oyster, the first water-resistant watch

- Charles Lindbergh makes the first non-stop solo flight across the Atlantic Ocean in *The Spirit of St Louis*, a specially adapted Ryan M-2 mail plane

- *The Jazz Singer*, starring Al Jolson, marks the start of the era of talking pictures

- Richard Drew invents Scotch adhesive tape

▶ The first Rolex Oyster

▼ Al Jolson and May McAvoy in *The Jazz Singer*

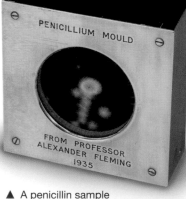

1929

EVENTS

- Black Thursday on the New York Stock Exchange marks the start of the Great Depression (1929)
- The Geneva Conventions on the treatment of prisoners of war and the sick and wounded are signed (1929)
- Leon Trotsky is expelled from the Soviet Union by Stalin (1929)
- Mahatma Gandhi launches his civil disobedience campaign against British rule in India (1930)

INVENTIONS

- Daniel Carasso builds a factory in France to produce yogurt on an industrial scale under the Danone trademark

- Donald Duncan, an American entrepreneur, buys up the rights to the bandalore, a traditional Filipino children's toy, from Pedro Flores, who relaunched it as the yo-yo

- The first in-flight refuelling of one aircraft from another takes place in the USA

- The BBC broadcasts its first television programmes

- Bell Laboratories in the USA produce the first quartz clock

- Edwin Hubble, one of the founders of modern astronomy, proves that the universe contains other galaxies besides our own and demonstrates that the galaxies are moving away from one another in an expanding universe, thereby providing powerful evidence for the Big Bang theory

- Alexander Fleming publishes the results of his work on penicillin

- Clyde W Tombaugh discovers Pluto, then thought to be the furthermost planet of the Solar System but now reclassified as a dwarf planet

- A Danish professor named Haxthausen takes infrared photographs of human skin, opening the way for thermal imaging

▲ A penicillin sample from 1935

1931

- The Statute of Westminster recognises the independence of the dominions of the British Empire (1931)
- Japanese forces invade Manchuria (1931)
- António Salazar establishes dictatorial rule in Portugal (1932)

- The first practical electric organs go into production in France

- Three Californians – Paul Barth, George Beauchamp and Adolph Rickenbacker – devise and market the electric guitar

- A US doctor named Earle Haar invents the tampon

- Long-playing records are first introduced in the USA

- Ernst Ruska, a German student engineer, invents the electron microscope

- Motor manufacturer André Citroën dispatches the Central Asian Mission across the Asian landmass

- The Owens-Illinois Glass Company starts marketing glass fibre, a product originally invented in 1836 by Ignace Dubus-Bonnel of France

- A US company called the Galvin Manufacturing Corporation markets the first car radio, soon changing its name to Motorola

- Newly discovered Freon gases help to make air conditioning more efficient

▼ Citroën's Central Asian Mission

1933

- President Franklin D Roosevelt launches the New Deal, to help rejuvenate the US economy (1933)
- Adolf Hitler becomes Chancellor of Germany (1933)
- The assassination of Sergei Kirov, secretary of the Loningrad branch of tho Soviot Communist Party, triggers the Stalinist purges (1934)

- Louis de Corlieu, a French naval officer, invents the first swimfins for divers

- Two US astronomers, Walter Baade and Fritz Zwicky, coin the word 'supernova' to describe a special category of extremely bright stars, which they correctly identify as being in the final stages of their lifecycle

- The first cathode-ray tube television sets go on sale

- A US game designer, Harry Williams, makes use of solenoids to transform existing pin games into electrically powered pinball

◄ An early car radio

▼ A *Concours Lépine* contestant in 1935

1935

- Italian troops take control of Ethiopia (1935)
- Mao's communists reach the end of the Long March (1935)
- Abdication of Edward VIII (1936)
- Germany reoccupies the Rhineland (1936)
- Civil war breaks out in Spain (1936)
- The Arab Revolt gets under way in Palestine (1936)

- The US geologist Charles Richter devises a comparative scale for measuring the magnitude of earthquakes

- While building careers as classical musicians, Americans Leopold Godowsky and Leopold Mannes invent a high-quality, easy-to-use colour film that will be marketed as Kodachrome

- The US chemist Edward Kendall and his team isolate cortisone

- George Gallup founds the American Institute of Public Opinion, which will systematise the use of opinion polls

- The world's first coin-operated parking meters are installed in Oklahoma City, USA

- Robert Watson-Watt develops a system of pulse radar that will play a crucial part in Britain's air defences in the Second World War

- Frédéric and Irène Joliot-Curie, the daughter of Pierre and Marie Curie, win the Nobel prize for physics for work on radioactivity, involving research on positrons and the discovery of artificial radioactivity

- Heinrich Focke successfully tests a functioning helicopter, the Fw 61

1937

- Getulio Vargas stages a coup in Brazil to instal the authoritarian Estado Novo regime (1937)
- In a mark of destruction to come, German bombers pound the Basque town of Guernica during the Spanish Civil War (1937)
- The Chinese city of Nanking falls to the Japanese (1937)

- Vladimir Demikhov transplants the first artificial heart into a dog, which survives for five and a half hours

- Pierre Debroutelle develops an inflatable boat while working for the French Zodiac company

- In the USA two employees of Standard Oil, Robert Thomas and William Parks, create butyl rubber, a malleable synthetic material that proves almost impermeable to gas

- Chester Carlson, a US patents clerk, develops a process he calls electrophotography, later known as photocopying

▼ Camping enjoys a boost between the wars as more and more people take to the great outdoors

◀ Igor Sikorsky tests the VS-300 helicopter

1938

- Germany annexes Austria in the Anschluss (1938)
- The Munich Agreement, signed by Germany, Britain, France and Italy, permits Germany to annex the Sudetenland (1938)
- King Carol II ends parliamentary rule in Romania (1938)
- Jews in Germany suffer the violent destruction of their lives and property in *Kristallnacht* (1938)

- The Volkswagen Beetle makes its first appearance as the KdF-Wagen

- The US monthly *Action Comics* publishes the first adventure of Superman, a character devised by Jerry Siegel and Joe Schuster

- South African fishermen catch a coelacanth, a fish belonging to a species thought extinct for 70 million years

- The US DuPont corporation takes out a patent on nylon, an artificial fibre developed by its employee, Wallace Carothers

- The first aeroplane to be equipped with a pressurised cabin is successfully tested in the USA

- Dr Rolla N Harger of Indiana University invents the Drunk-o-meter, forerunner of the breathalyser

- The Swiss firm Nestlé launches Nescafé, the first soluble instant coffee

1939

- Germany and USSR sign a non-aggression pact (1939)
- German invasion of Poland on 1 September makes war with Britain and France, Poland's allies, inevitable (1939)
- Fall of France and Allies retreat from Dunkirk (1940)
- Japanese aircraft attack the American naval base of Pearl Harbor on Hawaii, bringing the USA into the war (1941)

- Paul Hermann Müller discovers the insecticidal properties of DDT while working for the Swiss firm Geigy

- In Germany the first jet plane, a Heinkel He 178, makes a successful test flight

- Frédéric Joliot-Curie, Hans Halban and Lew Kowarski take out a patent on the production of energy from uranium by chain reaction – an important step on the path to nuclear power

- Four teenagers in the Dordogne region of France discover the Lascaux cave system, containing paintings dating back more than 20,000 years

- Alan Turing proposes his theoretical machine, prefiguring computers

▶ *Action Comics* launches its new hero: Superman

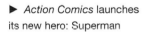
▼ A woman wastes no time in putting on her new nylon stockings

▼ German soldiers use an Enigma machine to encrypt a message

1943

EVENTS

- Soviet forces win the Battle of Stalingrad (1943)
- German troops destroy the Warsaw ghetto (1943)
- D-Day: Allied troops land in Normandy (1944)
- Liberation of France (1944)
- Fall of Berlin (1945)
- Atomic bombs dropped on Hiroshima and Nagasaki (1945)
- The United Nations Organisation is created (1945)

INVENTIONS

- American scientists identify fast-flowing air currents in the Earth's atmosphere as jet streams

- Jacques-Yves Cousteau and Émile Gagnan invent the aqualung, the breathing kit used for scuba diving

- Research on the temperature of animals undertaken by two US zoologists, Raymond Bridgman Cowles and Charles Mitchill Bogert, leads them to classify creatures as endotherms or ectotherms rather than warm-blooded or cold-blooded

- The German military targets the V-1, the first cruise missile, and the V-2, the earliest ballistic missile, at Britain

- Percy Spencer, a US military engineer working on radar, develops the microwave oven

- The University of Pennsylvania publishes a report on the construction of an automatic variable calculator, soon to be called a computer

1946

- The Nuremberg trials pronounce judgment on Nazi war criminals (1946)
- The US government launches the Marshall Plan, a vast reconstruction programme designed to restore Europe's shattered economies (1947)
- The Indian Independence Act separates India into two independent states: India and Pakistan (1947)

- The Italian firm Piaggio takes out a patent on the Vespa, the first modern scooter

- In the USA Earl Tupper creates the Tupperware brand of lightweight, non-breakable polyethylene kitchenware

- The American meteorologist Vincent Schaefer develops a technique for seeding clouds to create rain or snow

- In Italy Achille Gaggia invents the espresso machine

- Willard Libby and a team at the University of Chicago develop the carbon-14 technique for dating ancient objects

- Edwin McMillan, a US physicist, develops the electron synchrotron, an almost perfect particle accelerator

- Two employees of Bell Laboratories invent the transistor, laying the foundations of modern electronics

- Hungarian-born Dennis Gabor invents holography while working for British Thomson-Houston at Rugby in Warwickshire

- A US Air Force C-53 makes a transatlantic flight on autopilot

- In the USA Walter Morrison makes a plastic flying disc that he will later name the Frisbee

▼ The first atomic bomb test at Alamogordo, New Mexico

▲ Cousteau and Gagnan's scuba-diving kit

▶ Prototype of the V-2 ballistic missile

1948

- State of Israel founded (1948)
- The Communist Party seizes power in Czechoslovakia (1948)
- Allied powers organise the round-the-clock Berlin Airlift to counter a Soviet blockade of the city (1948-9)
- Mao Zedong proclaims the People's Republic of China in Beijing (1949)
- NATO is founded (1949)

- Edwin Land develops a revolutionary new camera that develops its own pictures: the Polaroid

- The Big Bang theory of the birth of the universe becomes generally accepted

- Auguste Piccard of Switzerland invents the bathyscaphe deep-sea submersible

- The world's first McDonalds restaurant opens in San Bernardino, California

- James Brunot markets *Scrabble*, originally devised a decade earlier by a US architect named Alfred Mosher Butts

- Two American friends, Bernard Silver and Norman Woodland, take out a patent on bar codes

1950

- Outbreak of the Korean War (1950)
- The USA snub's Mao's People's Republic of China by recognising the Nationalist government of the island of Taiwan as legitimate rulers of all China (1951)
- France, Germany, Italy, the Netherlands, Belgium and Luxembourg sign the Treaty of Paris setting up the European Coal and Steel Community, precursor of the European Union (1951)

- Frank McNamara and associates establish the Diners Club, introducing the use of credit cards

- Glass-ceramic materials are discovered by accident; unlike other forms of glass, they can tolerate rapid temperature changes and are shock-resistant

- A French doctor named Alain Bombard crosses the Atlantic in an inflatable boat

- Work undertaken by a Canadian heart surgeon, William Bigelow, leads to the development of the artificial pacemaker

- Sonographic techniques exploiting ultrasound are developed as a new method of medical imaging

- US scientists develop the hydrogen bomb; the first test destroys the Pacific island of Elugelab in the Enewetak Atoll, local residents having been compulsorily evacuated

▲ Jet stream above the Red Sea

▶ An early microwave oven

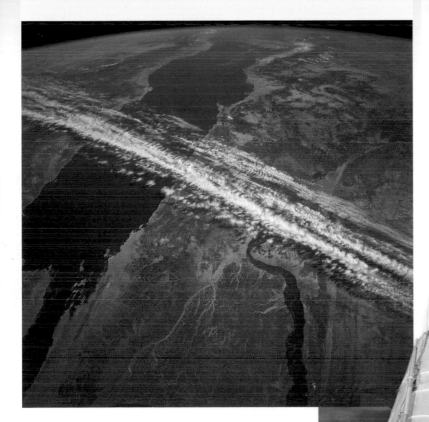

▶ Alain Bombard in his Zodiac inflatable

Index

Page numbers in *italics* refer to captions.

Picture credits

THE ADVENTURE OF DISCOVERIES AND INVENTIONS
Into the Nuclear Age – 1925 to 1945
Published in 2011 in the United Kingdom by Vivat Direct Limited
(t/a Reader's Digest), 157 Edgware Road, London W2 2HR

Into the Nuclear Age – 1925 to 1945 is owned and under licence from
The Reader's Digest Association, Inc. All rights reserved.

Adapted from *Les Inventions au Temps du Nucléaire*, part of a series entitled
L'ÉPOPÉE DES DÉCOUVERTES ET DES INVENTIONS, created in France by
BOOKMAKER and first published by Sélection du Reader's Digest, Paris, in 2010.

Translated from French by Tony Allan

PROJECT TEAM
Series editor Christine Noble
Art editor Julie Bennett
Designer Martin Bennett
Consultant Ruth Binney
Proofreader Ron Pankhurst
Indexer Marie Lorimer

Colour origination FMG, London
Printed and bound in China

VIVAT DIRECT
Editorial director Julian Browne
Art director Anne-Marie Bulat
Managing editor Nina Hathway
Picture resource manager Sarah Stewart-Richardson
Technical account manager Dean Russell
Product production manager Claudette Bramble
Production controller Sandra Fuller

We are committed both to the quality of our products and the service we provide to our
customers. We value your comments, so please feel free to contact us on 0871 35110000
or via our website at **www.readersdigest.co.uk**

If you have any comments or suggestions about the content of our books, you can
email us at **gbeditorial@readersdigest.co.uk**

CONCEPT CODE: FR0104/IC/S
BOOK CODE: 642-010 UP0000-1
ISBN: 978-0-276-44522-4